Practical Railway Engineering

Practical Railway Engineering

Edited by Marshall Roy

MURPHY & MOORE
www.murphy-moorepublishing.com

Murphy & Moore Publishing,
1 Rockefeller Plaza,
New York City, NY 10020, USA

ISBN: 978-1-63987-446-0

Cataloging-in-Publication Data

Practical railway engineering / edited by Marshall Roy.
p. cm.
Includes bibliographical references and index.
ISBN 978-1-63987-446-0
1. Railroad engineering. 2. Transportation engineering.
3. Railroads. I. Roy, Marshall.
TF145 .P73 2022
625.1--dc23

For information on all Murphy & Moore Publications
visit our website at www.murphy-moorepublishing.com

MURPHY & MOORE

Contents

Preface

Railway engineering is a multidisciplinary field in transportation engineering. It is concerned with the design, construction and operation of all types of rail transport systems. Railway engineering comprises mechanical design skills and knowledge of propulsion systems to design train vessels. Railway engineering includes civil engineering, computer engineering, electrical engineering, mechanical engineering, industrial engineering and production engineering. Command, control and railway engineering, energy electrification, overhead contact system, light rail systems, on-track plant, train control systems, railway signalling and systems integration are some of the sub-fields of railway engineering. This book attempts to understand the multiple branches that fall under the discipline of railway engineering and how such concepts have practical applications. The various studies that are constantly contributing towards advancing technologies and evolution of this field are examined in detail. This book is a vital tool for all researching or studying railway engineering as it gives incredible insights into emerging trends and concepts.

This book is a comprehensive compilation of works of different researchers from varied parts of the world. It includes valuable experiences of the researchers with the sole objective of providing the readers (learners) with a proper knowledge of the concerned field. This book will be beneficial in evoking inspiration and enhancing the knowledge of the interested readers.

In the end, I would like to extend my heartiest thanks to the authors who worked with great determination on their chapters. I also appreciate the publisher's support in the course of the book. I would also like to deeply acknowledge my family who stood by me as a source of inspiration during the project.

Editor

1

Advanced Train Positioning/ Communication System

Fouzia Elbahhar and Marc Heddebaut

Abstract

In the past, in order to ensure train positioning as well as ground-to-train information exchange, railways have adopted various technologies. Over time, each new generation of equipment enriched the global information exchange but, as a consequence, necessitated higher data rate transfers. For the positioning functionality, the existing localisation systems are still limited, since most of them require an infrastructure installation with constraints such as laying equipment between the rails or having high database maintenance requirements and computational costs. Moreover, some of them accumulate errors (odometers and inertial sensors) or offer limited coverage in shadowed areas (GNSS, etc.). Currently, in railway applications, a widely used localization system is based on proprio-ceptive sensors embarked in the train. This on-board system is coupled to the use of balises located at ground between the rails. These balises are kilometre markers. They are used to compensate for the drift of the localization information computed using the proprioceptive sensors alone, when the train moves. The balises provide absolute localization information whenever the train passes over them. They can also provide spot communication during the short period of time when trains are passing over them. In the first part of this chapter, techniques for achieving train positioning and data exchanges between trains and infrastructure are introduced. In the second part, a new balise is proposed. Particular attention is paid to the contribution of this new solution in terms of localization error and communication performances.

Keywords: railway balise, MISO-UWB, time reversal technique, localization error, physical layer performances

1. Introduction

This chapter is organised in two main parts. Sections 1–7 present railway requirements in terms of train tele-positioning and self-positioning systems. It also describes some corresponding generic technical solutions. Moreover, in Sections 5 and 6, the document focuses on the use of railway balises installed along the track. They are exploited for train positioning, as well as spot ground-to-train communications. Current existing technologies for these balises are then recalled, and potential technical limitations are deduced. In the second part of the chapter, in Section 8, a new balise using the Ultra-Wideband Impulse-Radio (UWB-IR) technique associated to the Time Reversal (TR) technique is introduced and extensively analysed. Then, Section 9 presents some experimental results characterising the operation of this balise. Finally, Section 10 draws a conclusion.

2. Train speed control

During the 1970s, a number of railway accidents came to recall that the safety of rail transport was not strictly guaranteed, and that there were, in particular but not only, some technical gaps to be filled. Accidents continued to occur in the 1980s, and experts noticed that stop signals, normally leading to the mandatory shutdown of trains, were often not respected by train drivers. These studies convinced railway operators that the operation of trains based solely on human control was not sufficient to ensure the desired level of safety for this mode of transport. It was therefore necessary to envisage the introduction of a speed control system that would complement and control the supervision carried out by drivers and, if necessary, imposed on them.

An effective train speed control requires access to several parameters. A vital parameter requires constant knowledge of the geographical location of all trains, also known as train positioning, circulating along the railway network. Train positioning, combined with train identification, as well as the knowledge of the characteristics of the track, makes it possible to generate, from a distant ground operator, speed instructions transmitted to the train, ranging from full speed allowed at this point of the railway network to an absolute stop. This allows safe circulation of all the trains circulating along the railway line. Of course, the safety requirements associated with the development of such a speed control and particularly train positioning of the trains arise are extremely high. Consequently, it imposes significant repercussions on the technologies likely to be used to ensure this technical function.

3. Safety integrity levels

Since in railway operation the lives of human beings are at stake, safety is of major concern for train speed control and therefore train positioning. In particular, dangerous failure modes must be reduced to an extremely low level. In railway operation, as in many other industrial operations, safety integrity levels (SILs) are used to determine the necessary target level of risk reduction. The IEC 61508 standard defines four SILs that correspond to different probability of failure per hour [1]. SIL 4 corresponds to the more critical situations and the lowest probability of failure per hour.

A SIL is determined using a number of quantitative factors that are combined with qualitative factors including hardware safety integrity and systematic safety integrity. As a consequence, SIL 4 must be achieved for speed control and for any train positioning hardware.

On the ground side, two major technologies are currently operational on most of the world's rail networks. They are based, on the one hand, on the use of track circuits and, on the other hand, on track balises providing positioning and ground-to-train spot radio communication. On the train side, an odometry processing unit exploiting different sensor technologies is used. Merging different technologies helps reducing common mode failures, as SIL 4 certification is mandatory for this equipment.

4. Track circuit, a safe train tele-positioning system

The track circuit is one of the very first electrical signalling systems used to ensure the safety of rail traffic. It makes it possible to detect the presence of at least one train present on a length of railway track called a fixed block and indirectly to ensure a safe distance between the trains running on the same line. Some also allow the transmission of speed and stop instructions to the train. Although its principle is very old, different manufacturers have developed many distinct technical achievements. Track circuit current use is almost universal on railway networks [2]. The general principle of operation of the track circuit is as follows: the track is physically divided into several fixed blocks from a few hundred metres to a few kilometres long. The length depends on the use of the line and the speeds of the trains circulating there. A signal, the characteristics of which are related to the type of track circuit used (frequency and amplitude), is emitted along the rails by a transmitter placed at one end of the block. If the receiver at the other end receives this signal, then the fixed block is declared unoccupied. This information is reported to the ground train speed control system. In the opposite case, if the train is located somewhere along the fixed block, the signal does not reach the receiver because the two rails are short-circuited by the axle of the train and the fixed block is declared occupied. This positioning information is determined from the ground, not by the train itself, e.g. tele-positioning, and reported to the distant ground-based train speed control system.

We notice that, for these track circuits, the two rails are used as a bifilar untwisted transmission line. Due to the use of steel rails directly placed on the ground, on ballast, this transmission line presents rapidly increasing losses, as the frequency increases. In practice, only low frequencies, up to a few kHz (1–2 kHz), make it possible to obtain communication ranges compatible with fixed block lengths up to 2–3 km [3].

From the point of view of train safety operation, it is therefore mandatory to detect the presence of effective trains on all the occupied fixed blocks and to be certain that these trains actually occupy the detected cantons, but not other fixed blocks. Depending on the type of separation used between the different fixed blocks, many types of track circuits exist based on the same general principle, but exploiting different technical approaches, not interoperable between them. In practice, separation between fixed blocks may be carried out by an insulating joint, which physically separates the various sections of the rails. With welded rails, low frequency, e.g. audio frequencies (AF) at 1–2 kHz can be used. An electrical joint based on a set of parallel inductances and capacitors acts as a filter preventing the passage of the low frequency signal from a fixed block to the following one. **Figure 1** presents such a filter used in a so-called UM71 AF jointless track circuit [4].

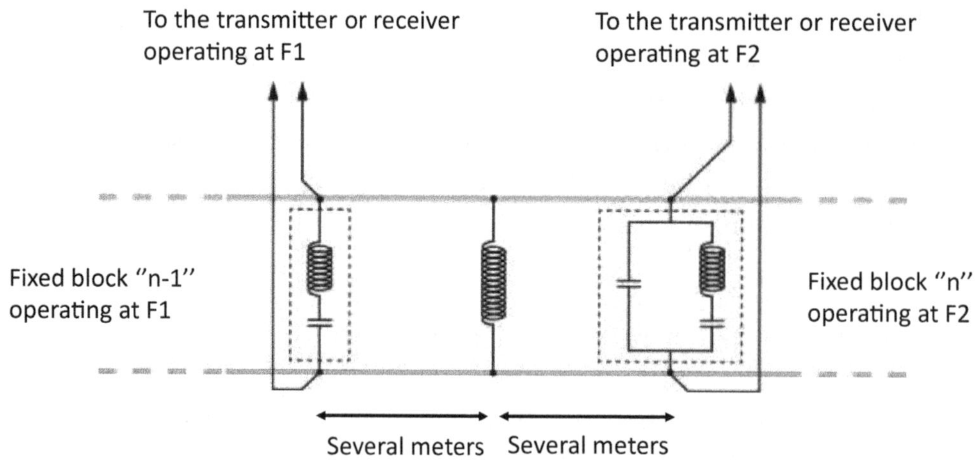

Figure 1. Filtering low frequency signals between two consecutive fixed blocks.

Fixed block noted n − 1 is travelled by a low-frequency F1 signal, which is heavily attenuated while passing through the filter and therefore almost undetectable in the following fixed block noted n. In the same way, fixed block n is travelled by a signal at frequency F2, which is also heavily attenuated while passing through the filter, so to be undetectable in the preceding fixed block n − 1. Therefore, all the consecutive fixed blocks can exploit the same alternation of frequencies F1 and F2. Frequencies F1 and F2 are chosen sufficiently low to enable a sufficient communication range and outside of potential harmonic signals coming from the electric train traction energy. They also must be sufficiently different to be effectively filtered. Of course, industrial choices can differ a bit, and technically, track circuits can be implemented differently from one network to another one, which also raises the problem of railway interoperability between different networks.

We deduce that track circuits provide a safe train detection of all the trains circulating along the railway level, with a spatial resolution limited to the length of the fixed block. Moreover, if the identification of the train is known by other technical means, then train tele-positioning, of course not very accurate since train positioning is only known somewhere along the fixed block, but safe, is provided.

5. Train odometry and balises

On the train side, measurements of the speed and position to deduce train self-positioning are continuously determined using an odometer exploiting different proprioceptive sensors. These sensors may consist of axle counters; they determine speed and covered distance knowing the circumference of the wheel [5]. To measure the speed and distance covered another way, microwave Doppler radars pointed to the ballast may also be used [6]. These Doppler radars exploit back scattering from the ballast, and their frequency can be optimised for the particular granulometry of the used ballast. **Figure 2** illustrates these two widely used sensor families.

Inertial sensors and global navigation satellite system (GNSS) sensors, among others are also used, are considered.

Figure 2. Proprioceptive sensors to feed the train odometer.

A measurement uncertainty and a progressive drift of the positioning accuracy are usually associated with each of these sensors. For example, due to wear, the circumference of the wheel is not perfectly stable and therefore not known perfectly. Consequently, over time, an increasing drift characterises the axle counter measurement. On its side, the Doppler measurement can be almost fully cancelled if water or wet snow lies between the rails. Indeed, this reduces or cancels the retro-diffusion effect. During the movement of the train, this results in temporary losses of information that affect the output information. Also, due to the low Doppler frequency generated, Doppler radars have limitations to measure very low speed, this drawback does not affect wheel counters. Considering these two sensor families, we verify that the use of very different sensor technologies helps to reduce common mode errors in order to build a global SIL 4 odometer.

Sensor drifts are compensated for balises regularly secured on the sleepers of the track. **Figure 3** shows such a balise installed between the rails. The trains embark a balise interrogator and reader located under the train body.

Figure 3. Example of balise mounted on the track.

These balises act as kilometre markers. They deliver to the passing trains their absolute positioning along the track. These data locally cancel the accumulated train odometer drift. After passing a balise, the train odometer restarts computing speed and position up to the next balise located on the track, at a distant where the accumulated drift is not yet problematic. These balises are also used in automated guided transport, for example, in order to calculate a precise station stop and to obtain the opening of the doors of the train exactly in front of the landing doors of stations.

These balises are often one of the latest, or even the latest, the last pieces of electronic equipment located at a very exposed and critical place, between the rails, on the track.

6. Speed control using balises

As described in the preceding section, balises are used as kilometre markers. They are installed regularly along the track, at accurately known locations, so that passing trains can read their absolute location on the fly. Moreover, these balises are also used to exchange data using spot ground-to-train radio communication. Balise-based speed control systems are installed in many railway networks. They provide automatic control of the speed and override of certain closed stop signals. In order to do this, the system uses a calculator on board the locomotives. The calculator takes into account both the information characterising the mobile and the data characterising the track and the location of the signals in order to determine a speed control curve that cannot be exceeded. This integrates the reaction time of each train equipment. Data are transmitted from the ground to the train through the balises. When a critical speed is reached, the train driver human machine interface sends an audible signal to warn the human operator of the need for urgent operation on his part to bring the speed back to a value acceptable to the system. If this action of the driver is not carried out quickly, the system will automatically stop the train by irreversible emergency braking, until it fully stops. The driver must then request from the ground controller a new driving authorization, after this total shutdown.

7. Current balise technology example

In Europe, balises are produced by a group of seven industrial companies: Alstom, Ansaldo STS, Bombardier, Invensys, Siemens, Sigma-Digitek and Thales, united within UNIFE, a federation of suppliers of railway equipment. This industrial grouping has defined and standardised eurobalise specifications and the associated test specification. The development of these specifications is driven by the European Railway Agency [7].

In the principle currently in play by eurobalises, the trains remotely power these beacons by magnetic coupling from a high-frequency 27 MHz, high-power signal emitted from an interrogator/beacon reader located under the train. These signals are emitted constantly as soon as the train is moving.

The magnetic coupling technology used by these balises was selected during the work carried out in the European project EURET 1.2, which was completed and published in 1996. Original Ericsson equipment, already a mature technology at that time, was compared to other techniques and has been selected and optimised in order to give the current eurobalise.

Benefits include interoperability over all the EU railway networks and efficient operation. Concerning the disadvantages, in addition to the electromagnetic pollution generated by the high power almost constant use of the train transmitter, it will be noted that this is the last indispensable equipment located at the track between the rails, which can generate constraints, in particular during the maintenance operations of the track. It is also an old analog technology, well below the current state of the art, using congested frequency bands and capable of generating and suffering electromagnetic disturbances in the vicinity.

A solution using satellite tracking techniques for trains is currently considered by several teams, but the requirements of railway signalling in terms of safety and reliability are strong and constraints linked to the physics of the problem leave room for doubt as to the emergence of a purely SIL 4 satellite-based track positioning solution for railway signalling. In particular, this is due to the wide disparity of environments likely to be encountered by trains in circulation.

8. Balises emerging technologies

Recently, railway balises have received renewed attention. Techniques considering state-of-the-art solutions rather than the old, currently used ones are being studied. In [6], fast-moving Radio Frequency IDentification (RFID) tags are studied in high-speed railway systems. However, authors identify issues, such as collision and insufficient reading time, and propose various ways to alleviate their effect in railway systems. Moreover, these new balises would remain located between the rails on the track. This proves to be constraining from the point of view of maintenance of the track, when it is necessary to add new ballast or to change rails, for examples.

A new generation balise using modern, energy-efficient, and green technologies is proposed in the following of this chapter. This balise is designed to operate from the side of the track and uses the ultra-wideband impulse radio (UWB-IR) radio technique. This technique possesses the intrinsic qualities of having a short-range, high communication capacity, of requiring low emitted power while providing precise relative location capability. Moreover, UWB technique has a low signal intercept capability by non-accredited users. Thus, there are strong arguments in favour of a future SIL 4 certification for a railway balise based on this technique.

This UWB pulse technique uses non-sinusoidal signals and transmits pulses of very short duration. They occupy an extended spectrum of at least 500 MHz bandwidth. Transmitters are of simple design and low consumption. These UWB transmissions are authorised worldwide in a specific power gauge [8]. Associated with the UWB-IR radio technique, a time reversal (TR) technique is also used [9]. The TR technique makes it possible to spatially and temporally focus an electromagnetic signal in a dispersive propagation medium. The principle of the TR of the waves is based on the invariance of the equation of propagation of waves by reversal of time. This invariance allows a wave to backtrack so that it can replay the "go" scene of its propagation but backwards. The operating principle of this new generation balise is presented in **Figure 4**. The conventional balise situated between the rails is removed and replaced by the new balise, installed on a pole, on the side of the track and a few meters away. This new balise focuses the radiofrequency energy coming from the pole transmitters to an area situated over the rails, right over the removed conventional balise location. Therefore, this new balise does not interfere anymore with track maintenance operations but still develops a maximum of radio frequency signal at this particular location over the rails. Several transmitters are coupled on the pole; three can be seen in **Figure 4** to get a multiple source transmitter. This insures transmitter redundancy as well as space focusing when correctly configured. One single receiver or train-balise reader is used located in front of the train. This configuration is usually denoted as a Multiple Input, three transmitters, Single Output, one receiver, MISO 3 × 1 system.

Table 1 establishes a comparison of performance between the communication and localization systems using a railway conventional balise "existing balise" and the expected performance of the presented TR-UWB balise.

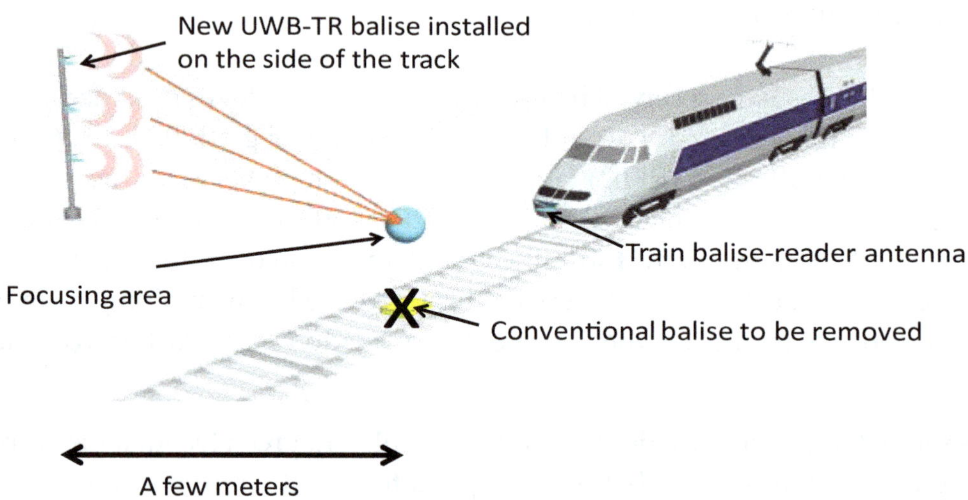

Figure 4. The new proposed railway balise.

	Conventional balise	**TR-UWB balise**
Operating frequency	27.095 MHz train to ground 4.5 MHz ground to train	3.1–10 GHz
Communication range	<1 m	10–100 m
Transmission rates	560 kbps	Potentially up to several hundred Mbps
Localization accuracy expected	20 cm	<10 cm

Table 1. Performance comparison between a conventional balise and the proposed balise.

8.1. Ultra-wideband radio technique

The UWB technology involves the transmission of very short pulses, typically having a time duration of 1 ns or less, therefore occupying a very wide frequency spectrum [10]. The use of this technology for our proposal is motivated by the following reasons:

- its potentially high transmission data rate, thanks to the very large frequency bandwidth used,
- its high resolution train location due to its very fine temporal resolution,
- its availability and robustness to multipath tunnel because of the large frequency bandwidth used,
- its low probability of detection due to a very low transmitted power; this property is necessary for safe and secure operation of the railway system,
- its ability to coexist with other radio systems thanks to low power spectral densities that do not require allocating a specific frequency band.

These reasons make UWB a novel, effective and complete solution for communication and localization in railways applications. Therefore, regulation bodies have considered the use of UWB in railway applications. **Table 2** provides some inputs regarding areas of operation of UWB systems [8].

However, the channel propagation effects, especially the lack of line of sight (LOS) between the transmitter and the receiver, and the presence of strong multipath are two significant sources of error for the localisation function. The introduction of UWB in wireless communication has brought improvements regarding these sources of errors [8]. Concerning the UWB technique, previous investigations have raised major issues, such as the complexity of the signal processing

Frequency (GHz)	Area of operation	Critical factors impairing system performance	Countermeasures compatible with limits and regulations
$3.11 < f < 4.8$ PSD < −53.3 dBm/MHz for unregistered UWB unlicensed mobile devices with 5% activity LDC PSD < −41.3 dBm/MHz for registered devices	Very short range (<10 m) short range (<50 m)	Multipath Broadband interferers (e.g. automotive UWB) Multipath + path loss Broadband interferers	Multiple UWB emitters onboard the train cars real-time processing Multiple "ground-based". fixed receivers Multiple UWB emitters at 3.1–4.8 GHz deployed as "ground-based" fixed references for real-time processing
$6 < f \leq 8.5$ PSD < −53.3 dBm/MHz for unregistered UWB mobile devices with 5% LDC	Very short range (<10 m)	Multipath	Multiple UWB emitters onboard the train cars. Real-time processing. Narrow-beam antenna

Note: At highest frequencies ($6 < f \leq 8.5$ GHz) very short-range (<10 m) applications only are affordable, due to the fact that fixed infrastructures made of UWB emitters are not allowed by the Electronic Communications Committee (ECC).

Table 2. Critical factors limiting the performance of UWB systems in railway environments.

at the reception [11, 12]. Therefore, UWB has been associated with TR [13, 14], especially in multi-users communication systems, in order to solve part of these problems and to report some of the noted complexity at the transmitter level.

8.2. Time reversal

Time reversal (TR) has been applied to acoustics and underwater systems [15]. It is closely related to the retro-directive array in microwave [16] and phase conjugation in optics. The first TR experiment using electromagnetic waves in the 2.45 GHz band was reported in [17]. This contribution suggests that the techniques developed for ultrasound might also be used for the study of electromagnetic case. It is an interesting challenge because in many real environments (buildings or cities), microwaves, using wavelengths between 5 and 30 cm, are scattered off by objects such as walls, desks, vehicles, and so on, which produces a multitude of paths from the transmitter to the receiver.

In such situation, a TR system should be able not only to compensate for the multipath effect but should also improve radio communication parameters, thanks to the many existing reflections/reverberations.

TR has two main characteristics, the temporal focusing and the spatial focusing; these are very beneficial to the UWB system [17, 18]. More recently, TR has also been studied for broadband radio communications, especially for UWB radio [13].

For our proposed TR-UWB system, we selected the second Gaussian derivative function as the signal shape to transmit [15]. Then, the channel impulse response (CIR) is measured between the transmitter (Tx) and the receiver (Rx), and the channel state information is loaded into Tx. The selected signal and the impulse response are then reversed in time and transmitted by Tx in the propagation channel, up to Rx. This process, represented in **Figure 5**, can be mathematically described by noting $s(t)$ the transmitted pulse, $h(t)$ the complex impulse response of the channel and $h^{*}(-t)$ the complex conjugate of the time reversed version of $h(t)$; $y(t)$ the received signal without TR and $y_{RT}(t)$, the received signal with TR at the receiver; one has:

$$y(t) = s(t) \otimes h(t) + n(t)$$

$$y_{RT}(t) = s(t) \otimes h^{*}(-t) \otimes h(t) + n(t)$$

$$heq(t) = h^{*}(-t) \otimes h(t)$$

Applied to our proposed railway balise, this TR general process becomes the following one. The local Channel State Information (CSI) between each balise transmitter and the distant area, where the energy must be focused is measured or computed a single time during the installation phase of the balise. Moreover, as trains will support the embarked balise reader, the CSI measurement is performed in presence of a metallic reflective mask representing the front of a train. It is situated at the distant area considered between the rails, where a maximum anticipated focusing effect is expected. This unique CSI is then loaded in the transmitter equipment to perform the TR operation. As long as the propagation environment remains unmodified, this initial CSI can be repetitively used by the balise. This information is then introduced as

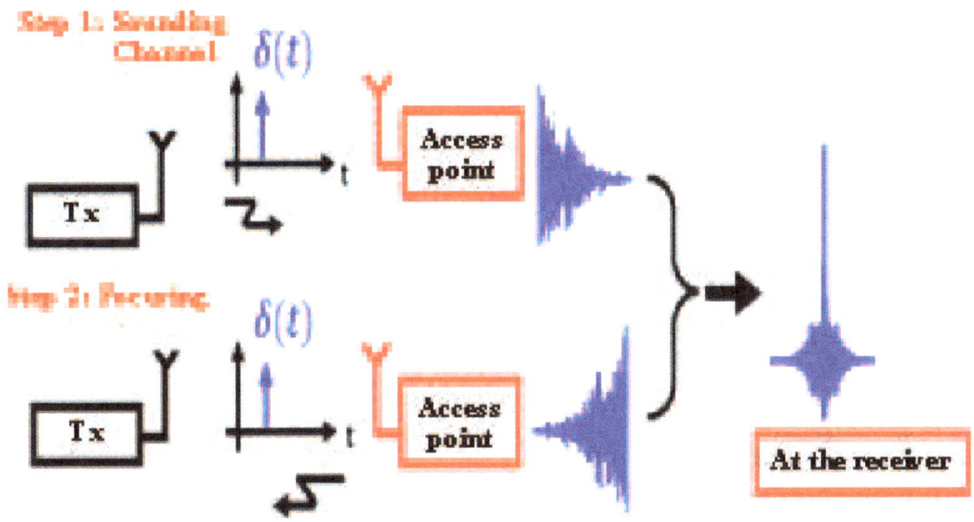

Figure 5. Principle of time reversal technique.

pre-filtering data in the different balise transmitters. Therefore, focusing is obtained in the required position, along the track, potentially improving the absolute localization process [14]. Moreover, using a single balise, focusing can be achieved successively toward different tracks, therefore, addressing different trains and delivering specific data to each train.

8.2.1. Parameters to evaluate TR effectiveness

The autocorrelation function is used to evaluate the temporal focusing (TF) and the spatial focusing (SF). To study TF, one can evaluate the Focusing Gain (FG), which is defined as the ratio of the spectrum power of strongest amplitude peak in TR received to the strongest peak received by a conventional UWB system. The focusing gain can be written as:

$$FG_{[dB]} = 20\log_{10}\left(\frac{\max(|y_{rt}(t)|)}{\max(|y(t)|)}\right) \tag{1}$$

Higher FG usually translates into higher communication range and higher precision of localization for a localization system as compared to a classical system.

The study of SF considering a simple transmitter to receiver configuration is performed the following way. The channel impulse response (CIR) of the intended receiver located in position p_0 is noted $h(p_0,t)$. The CIR of the unintended receiver located in position $p_{i(i\neq 0)}$ is noted $h(p_{i,\,i\neq 0})$. The equivalent CIR of the intended receiver is then given by:

$$heq(p_0,t) = h^*(p_0,-t) \otimes h(p_0,t) \tag{2}$$

While the equivalent impulse response of the unintended receiver is given by:

$$heq\left(p_{i,i\neq 0},t\right) = h^*(p_0,-t) \otimes h(p_i,t) \tag{3}$$

SF is then evaluated as the ratio of the strongest peak power received by the intended receiver to the strongest peak received by the unintended receiver. The SF parameter can be written as:

$$SF_{[dB]} = 20\log_{10}\left(\frac{\max(|heq(p_0,t)|)}{\max(|heq(p_1,t)|)}\right) \tag{4}$$

In a conventional railway environment, the use of a MISO 3 × 1 configuration is considered sufficient. We evaluated the contribution of TR in this configuration using the FG and SF. The study of SF and FG is performed for channel models exploiting the IEEE 802.15.3a channel model. This model is based on the Saleh Valenzuela formalism [8].

In the case of the IEEE 802.15.3a model, the characteristic values of Nakagami-m were explored, taking into account the number of antennas N_t.

Expression of the CIR is given by Eq. 5:

$$heq_{MISO}(t) = E\left\{\sum_{i=1}^{N_t} heq_i(t)\right\} \tag{5}$$

where

$heq_i(t) = \int_A \sum_{m=0}^{\infty} \alpha_{mi}^2 s_{mi}(\tau - t_{mi}) s_{mi}(\tau + t - t_{mi}) d\tau$

Then,

$$\begin{aligned} heq_{MISO}(t) &= E\left\{\int_A \sum_{i=1}^{N_t}\sum_{m=0}^{\infty} \alpha_{mi}^2 s_{mi}(\tau - t_{mi}) s_{mi}(\tau + t - t_{mi}) d\tau\right\} \\ heq_{MISO}(t) &= E\left\{\sum_{i=1}^{N_t}\sum_{m=0}^{\infty} \alpha_{mi}^2\right\} \Phi'_{si}(t) \\ heq_{MISO}(t) &= \sum_{i=1}^{N_t}\left[E\left\{\sum_{m=0}^{\infty} \alpha_{mi}^2\right\}\right] \Phi'_{si}(t) \end{aligned} \tag{6}$$

where $\Phi'_{si}(t) = \int_A s_i(\tau - t_i) s_i(\tau + t - t_i) d\tau$

The calculation of the average energy of the CIR in generic interval $W = [a, b]$ (a and b are arbitrary chosen) provides:

$$E\left\{\sum_{i\in I_W} \alpha_i^2\right\} = \int_W P_g(t)dt \tag{7}$$

where I_W is the random set containing the multipath components. The variance of the energy function of the CIR is given by:

$$Var\left\{\sum_{i\in I_W}\alpha_i^2\right\} = \int_W R_g(t)dt \tag{8}$$

where Rg(t) is the kurtosis of the delay profile. Exploiting Eqs. (7) and (8), the expression $heq_{MISO}(t)$ becomes:

$$heq_{MISO}(t) = E_g\sum_{i=1}^{N_t}\Phi'_{si}(t) \tag{9}$$

The corresponding ($PDP^{MISO}_{ULB-RT}(t)$) is given by:

$$PDP^{MISO}_{TR-UWB}(t) = E\left\{\left|heq_{MISO}(t)^{MISO}\right|2\ \right\} \tag{10}$$

After development, Eq. (10) becomes:

$$PDP^{MISO}_{TR-UWB}(t) = \sum_{i=1}^{N_t}\left\{E_g^2\Phi'_{si}2(t) + \frac{E_g^2}{2\tau_{rms}}^{*}\varpi\right\}$$

where $\varpi = \left[\left(1+\Psi_{\Phi'_{si}}(t)\right)c_{1i}\exp\left(-t/\tau_{rms}\right) + \left(1+\frac{1}{m'}\right)1\bar{\lambda}\Phi'_{si}2(t)\right]$ and,

$$c_{1i}\Psi_{\Phi'_{si}}(t) = \int_{A'}\phi_s(\xi+t)\phi_s(\xi-t)d\xi d\tau,$$

$\Psi_{\Phi'_{si}}(t)$ is the normalised autocorrelation of $\Phi'_{si}(t)$ $\Psi_{\Phi'_{si}}(0) = 1$.

The corresponding focusing gain is given by Eq. (11):

$$FG_{[dB]} = 10Log_{10}\left[\frac{E_g\sum_{i=1}^{N_t}\tau_{rms} + c_{1i} + c_2}{N_t}\right] \tag{11}$$

where $c_2 = \left(1+\frac{1}{m'}\right)\frac{1}{2\bar{\lambda}}$, m' represents the Nakagami-m value.

The values obtained on the evaluation of focusing gain for MISO 3×1 configuration are presented in **Table 3**. These values correspond to the different IEEE 802.15.3a Channel Model configurations known as CM1, CM2 and CM3 corresponding to different increasing channel complexities. We obtain that changing from CM1 to CM3, FG increases from 14.6 to 20.1 dB. Therefore, as expected, the focusing gain increases, when the channel complexity increases.

IEEE 802.15.3a channel model	CM1	CM2	CM3
$FG_{analytical}$ [dB]	14.4	16.1	19.9
$FG_{simulation}$ [dB]	14.6	16.2	20.1

Table 3. FG in analytical and simulation study, case of MISO 3 × 1 (IEEE 802.15.3a channel model).

8.3. Contribution of TR-UWB system in terms of communication performance

UWB transmission optimised through Time-Reversal (TR) technique [8] is investigated and proved to be advantageous in multipath environments. This solution allows increasing the Signal-to-Noise Ratio (SNR) and therefore the communication range. This reduces interference effects and receiver complexity, improves data rates and multiuser capacity and provides Low Probability of Interception (LPI) to non-intentional users.

Under these assumptions, our simulations propose to demonstrate the benefit of UWB-TR transmission in the tunnel, especially in terms of ISI and MUI reduction, as well as providing higher throughputs at extended communication ranges. Both Single Input Single Output (SISO) and Multiple Input Single Output (MISO) transmissions were tested for both the conventional transmission mode (without TR) and the TR one. To consider and evaluate the performance of UWB-TR communication system in a very constraint railway environment, we introduced a channel model dedicated to tunnel environments. Moreover, to cope with this difficult environment, up to a MISO 4 × 1 is considered in this environment. We used a geometrical ray model simulating a tunnel. Based on geometric optics and the ray-tracing techniques, a straight 8 m wide and 6 m high, rectangular, infinite tunnel is simulated [8]. Uniform cross-section and lossy, smooth walls are assumed. Regarding this geometry and tunnel cross section, very large relative to our studied wavelengths (10 to 3 cm), this deterministic approach seems accurate and fast to implement and run. The model is exploited in the frequency domain, over the 3.1–10.6 GHz FCC2 authorised band. The transmitter is located at a particular place along the main axis of the tunnel, at different lateral positions (P1–P4), according to the parameter to estimate. The receiver moves along the main axis of the tunnel from a reference distance (1 m), then starting from 5 m up to a given distance noted dn, using a 5-m step. The averaged and normalised power delay profile (PDP) are given in **Figure 6**. We evaluated temporal compression for three different Tx-Rx distances, d = 5, 50 and, finally, 100 m. Three scenarios have been tested: SISO without TR, SISO with TR and MISO-TR with four transmitting antennas (NTx = 4). For SISO configurations, both Tx and Rx antennas were centred on the tunnel cross section. For MISO situations, as depicted in **Figure 7**, Rx antenna is centred, Tx antennas are positioned at the corners of a 50-cm square, one corner being the centre of the tunnel.

To evaluate the temporal focusing effectiveness, we computed the RMS (root-mean-square) delay spread using non-TR and TR configurations (see **Table 4**). As demonstrated in [14] for residential indoor environments case, the predicted delay spread in the tunnel environment for the SISO-TR case is not reduced. However, SISO-TR results in a highly peaked main lobe signal above the temporal side lobes. In the case of MISO-TR, temporal focusing is clearly improved. Results also show that using the highest number of transmitting antennas produces

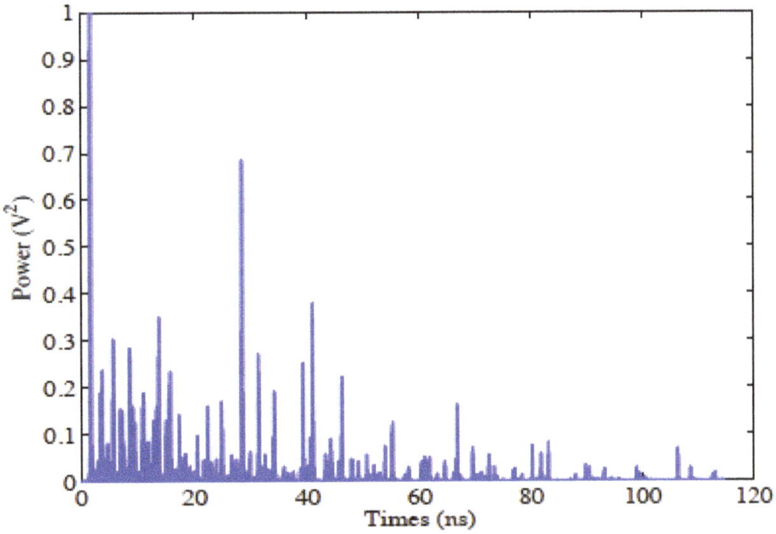

Figure 6. Averaged and normalised power delay profile (PDP).

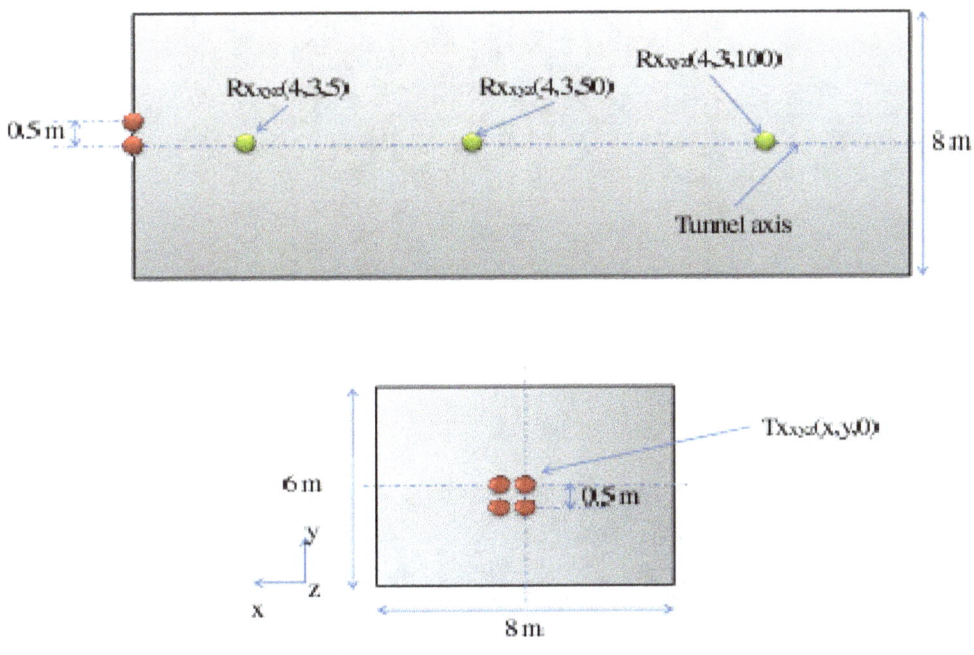

Figure 7. Time focusing simulation setup.

Distance	5 m	50 m	100 m
Without time reversal	31.55 ns	13.93 ns	8.20 ns
Time reversal SISO	32.55 ns	13.95 ns	8.42 ns
TR MISO NTx = 4	16.31 ns	10.62 ns	7.47 ns

Table 4. RMS delay spread comparison.

the better temporal compression, especially at the shortest Tx-Rx distances (by about a factor of 0.5). This can be explained by the scattering effect, more important at shortest distances than at farther distances. Consequently, the transmitter correlations are lower, resulting in a better temporal compression.

To evaluate the spatial focusing, two configurations were investigated for a Tx-Rx separation of d = 5 m: SISO-TR and MISO-TR (with NTx = 2). **Figure 8** illustrates power normalised to peak received in the area without TR. We observe that all users in this area receive approximately the same power level. This means that the communication can be intercepted by any receiver, and, in our case, the MUI problem can really affect the multiple access performance. By using TR (**Figure 9**), the power is decaying rapidly when moving away from the target area.

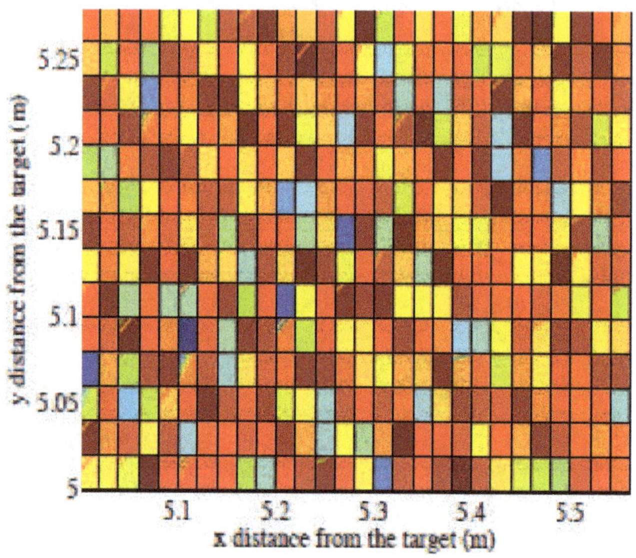

Figure 8. Power spatial repartition in the reception area.

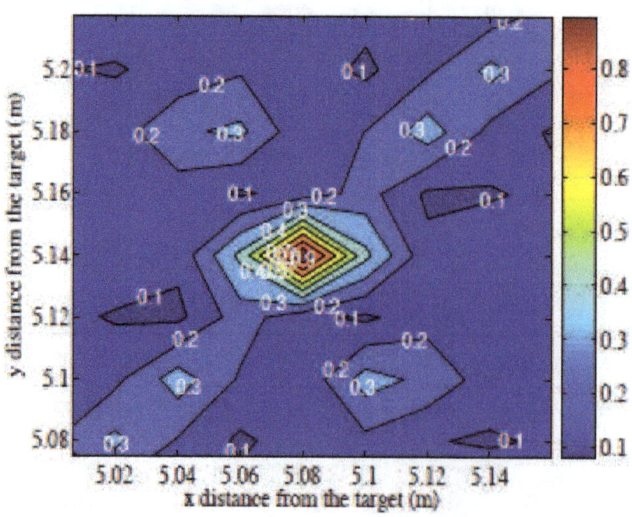

Figure 9. Spatial focusing gain in the SISO-TR case.

8.4. Contribution of TR-UWB system in terms of localization accuracy

The goal of this section is to estimate the contribution of TR in terms of precision of localization as a function of the propagation environment complexity. Using three base stations, we determine the position of a mobile in a 2D plane. To locate the mobile, each base station sends its own signal (recorded and reversed in time in the case of TR). As indicated before, the UWB-IR signals are built using the second-derivative Gaussian pulse. They are modulated by an antipodal modulation and coded by a Gold code, one code per base station. Processing is performed at the mobile (Rx) to determine its position relative to the base stations. The distance error is given by the difference between the calculated and the actual position of the mobile. The mobile receives the signals from each base station and performs an adequate signal processing to determine its position, relative to the base stations. Using the TDOA technique, the signal received at the mobile is processed to retrieve the position of the latter [19]. The Chan algorithm is used to solve the non-linear equations imposed by the TDOA technique. For the comparative study between the conventional localization system UWB-IR and the proposed TR_UWB system, we first use the UWB- IR case to locate the mobile, and then, the obtained information on the position of the mobile is used as reference for locating with the TR-UWB system. Our comparison is based on the computation of the Root Mean Square Error (RMSE) of localization between the conventional UWB-IR system and the proposed TR-UWB-IR system. We consider the IEEE 802.15.3a channel model (CM3). An AWGN is also injected into the channel. These results show that the combination of UWB and TR allows for a more accurate localization to be obtained that could be in line with the decimetre necessary level of precision required by the railway beacon application (**Figure 10**).

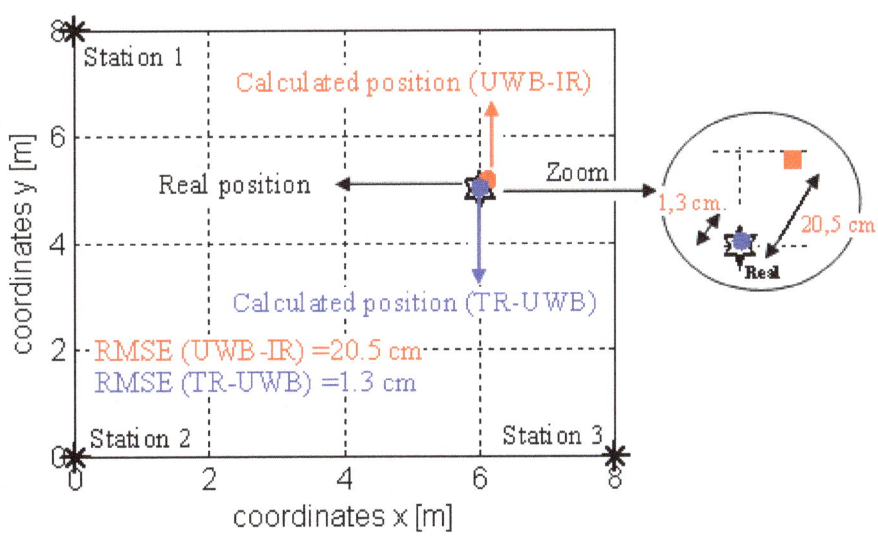

Figure 10. RMSE evaluation for conventional UWB-IR and TR-UWB (2D positioning system).

9. Experimental validation

The purpose of this experimental validation is to assess the impact of environmental complexity on performance related to temporal/spatial focusing and positioning error and to compare these conclusions to our preceding simulation results.

9.1. Experimental setup

An Arbitrary Waveform Generator (AWG) associated with a fast sampling oscilloscope (TDS) is used. The pulses generated by the AWG are radiated using wideband horn antennas. Similar antennas are used for receiving; their outputs are connected to the TDS ports through low noise amplifiers (LNAs). A portable computer is used to process the signals and store the results. The setup is installed in an anechoic chamber to control the radio channel. In this environment, metallic reflectors are introduced to create, on demand, different configurations of multipath. The dimensions of the anechoic chamber we used are 7 × 7 × 3 m. It is operating from 100 MHz to 10 GHz. Two antenna cases are considered: SISO, MISO 3 × 1 cases. For each type of case, three configurations are considered: without addition of metal reflectors; a single aluminium plate (2 m × 1 m) is introduced as a reflector between transmitters and receiver to generate a first multipath configuration; then, four plates are present to maximise the propagation environment complexity. The objective of these tests is to evaluate the focusing gain (FG), on the one hand, as the complexity of the propagation channel increases and, on the other hand, as the antenna configuration evolves from SISO to MISO 3 × 1. For the SISO configuration, the distance between the transmitting antenna and the receiving antenna is set to 5 m, which is representative of an in-situ railway balise application. MISO 3 × 1 configuration corresponds to the addition of a third transmitting antenna located at a distance of 4 m between Tx3 and Rx. In a first step, for each selected configuration, a pulse is transmitted using the AWG; the received signal is acquired by the TDS and then returned temporally. In the cases of SISO and MISO 3 × 1, each Tx re-emits its corresponding reversed in time signal. To assess this temporal focusing, we calculate the focusing gain (FG) obtained in each case. The overall results are grouped in **Table 5**. We obtain that, for each of the configuration, the focusing gain increases with the number of reflectors introduced. Moreover, by comparing the values of FG for these configurations, we find that FG increases from SISO to MISO configuration. To evaluate SF, we consider the scenario using three reflector plates. For the two SISO and MISO 3 × 1 configurations, the receiver is moved by 10 cm from its initial position. We note the initial position p0 and p1 and then the resulting ones after displacement. **Figure 11** represents an illustration of the SISO experimentation. The signal previously reversed in time at position p0 is transmitted, and this signal is now received at position p1, where the used CSI is no more optimal. We then evaluate SF obtained at position p0, compared to SF obtained at position p1, using our two configurations and three reflector scenarios. The results are summarised in **Table 6**. We obtain that SF values increase with the number of reflectors but also as a function of the number of transmitting antennas. This confirms the results obtained in theory and verified by simulation (**Figure 12**).

Configuration	**SISO**	**MISO 3 × 1**
$FG_{[dB]}$ **(without reflector)**	1.0	1.7
$FG_{[dB]}$ **(1 reflector plate)**	2.2	5.9
$FG_{[dB]}$ **(4 reflector plates)**	6.1	12.8

Table 5. Focusing gain (FG) according to the number of reflector plates inserted in the propagation environment (case of SISO, and MISO 3 × 1 configurations).

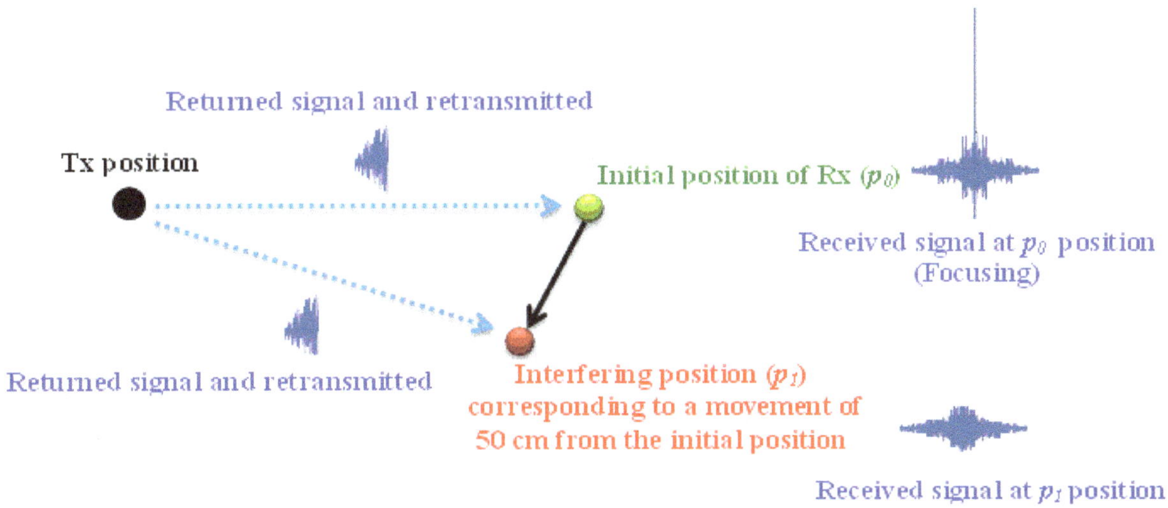

Figure 11. Principle of SF measurement.

Configuration	SISO	MISO 3 × 1
$SF_{[dB]}$ **(without reflector)**	2.3	3.1
$SF_{[dB]}$ **(1 reflector plate)**	5.8	8.1
$SF_{[dB]}$ **(4 reflector plates)**	12.3	16.4

Table 6. SF according to the number of reflector plates inserted in the propagation environment.

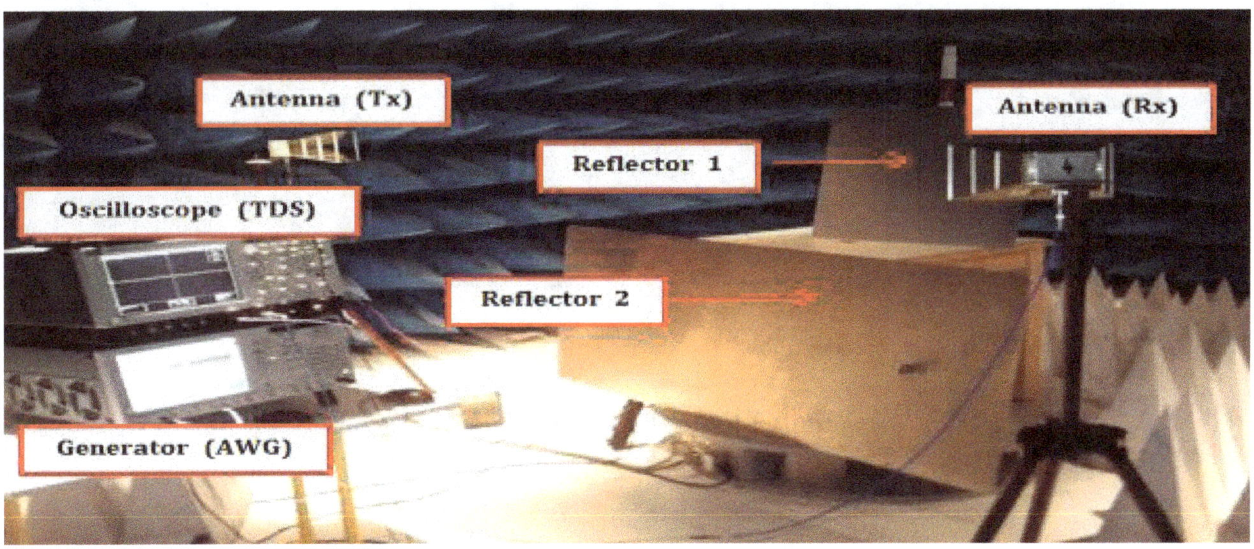

Figure 12. Implementation of the selected configurations (reflector plates in SISO).

9.2. Contribution of TR-UWB system in terms of communication performance

In the anechoic chamber, we considered five different reflector configurations. Three series of acquisition were performed. In each case, a sequence of seven pulses is sent. After processing, we determine the position of the mobile. **Figure 13** shows the corresponding position errors.

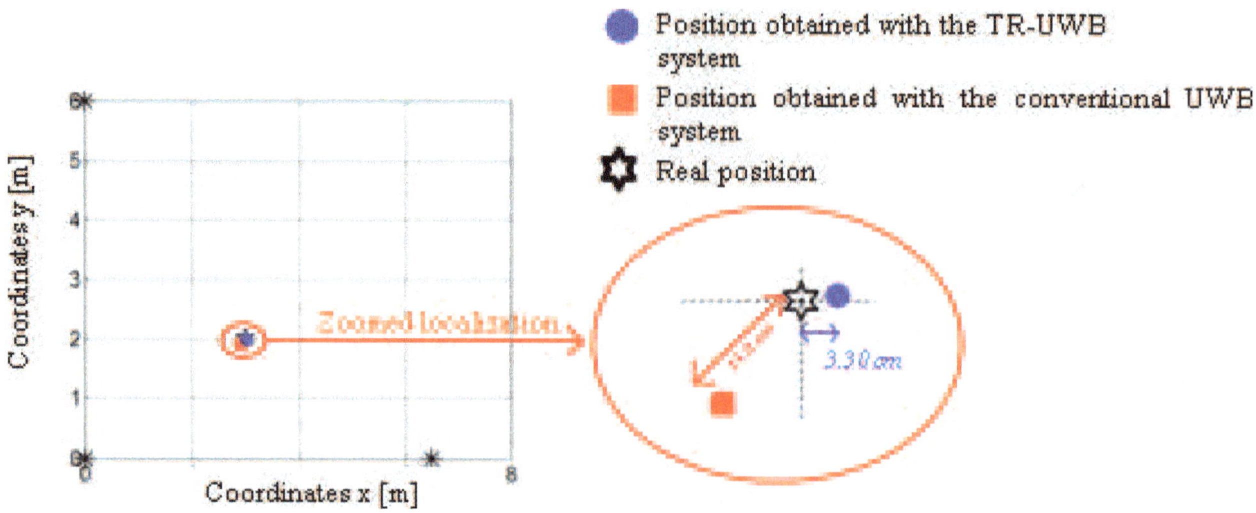

Figure 13. Conventional UWB and TR-UWB localization systems (scenario with three reflector plates in an anechoic chamber).

Figure 14. Implementation of tests configuration.

With the conventional UWB localization system, we get an error of 11.0 cm, decreasing to 3.3 cm in the case of TR-UWB (**Figure 14**).

10. Conclusion

As mentioned in Section 3, railway beacons are safety critical pieces of equipment. Therefore, a final industrial product will have to comply with SIL 4 requirements. To achieve this high level of integrity, one major goal consists in demonstrating that no single failure of this piece of equipment would lead to a railway safety critical situation. For such a demonstration, let us examine the advantages that are provided by the proposed association of UWB and time reversal techniques. The UWB radio technique alone already provides some interesting advantages that will intrinsically facilitate a future SIL 4 demonstration. First, an UWB-IR radio transmitter contains a limited number of components. Therefore, a safety demonstration will be easier to be realised than for a more sophisticated conventional radio transmitter. Complementary major advantages related to the use of the UWB technique are its high throughput capacity due to its wide frequency range and its availability and robustness to multiple paths, thanks to the large occupied bandwidth. The coexistence of the UWB signals with other radio systems due to the low involved power spectral densities is also interesting to note; this usually implies that there is need for a dedicated frequency allocation. Moreover, in a railway environment characterised by large radio interference, the intrinsic resistance of UWB signals to narrow band or even wide band radio interference is also very beneficial. Finally, due to its low power and large spectrum used, the low probability of interception of UWB transmitted signals by non-accredited users is also quite interesting. Indeed, this property will be in more demand for a secure operation of the transport system. As demonstrated in the previous sections, coupling the time reversal precoding technique to the UWB radio technique also improves some of these initial advantages in terms of higher throughput, accuracy of location etc. Moreover, since a MISO radio scheme is used, the failure of one transmitter among the different ones used for time reversal will reduce the spatial and temporal focalisation levels of performance, but the critical radio link with the train will still be maintained. Concerning jamming, spoofing and meaconing attacks that are becoming more critical in satellite systems, civil location and time assurance, spatial focusing delivers significant improvements. Indeed if valid beacon signals can only be received in a very limited zone or time window while the train is running, this property can be used to reject signals that are coming from non-valid distant transmitters, which do not present these focusing characteristics.

Acknowledgements

This present research work has been supported by SAFER-LC H2020 European Project, InterCor Project and ELSAT 2020. ELSAT 2020 project is co-financed by the European Union with the European Regional Development Fund, the French state and the Hauts de France Region Council.

Author details

Fouzia Elbahhar[1,2]* and Marc Heddebaut[1,2]

*Address all correspondence to: fouzia.boukour@ifsttar.fr

1 Univ. Lille Nord de France, Lille, France

2 IFSTTAR, LEOST, Villeneuve d'Ascq, France

References

[1] Kirkwood D, Tibbs B. Developments in SIL determination. IET Computing & Control Engineering Journal. June-July 2005;**16**(3):21-27

[2] Frederic Howard L. Track-circuit signaling on electrified roads. Proceedings of the American Institute of Electrical Engineers. Aug 1907;**26**(8):1277-1292

[3] Hill RJ. Optimal construction of synchronizable coding for railway track circuit data transmission. IEEE Transactions on Vehicular Technology; **39**(4):390-399

[4] Ogunsola A, Mariscotti A. Electromagnetic Compatibility in Railways: Analysis and Management, Lecture Notes in Electrical Engineering. Verlag Berlin Heidelberg: Ed. Springer; 2013. pp. 119-125

[5] Transportation Research Board (TRB). Track Design Handbook for Light Rail Transit, TCRP 105. 2nd ed. pp. 10-11

[6] Zhang X, Tentzeris M. Applications of fast-moving RFID tags in high-speed railway systems. International Journal of Engineering Business Management. 2011;**3**(1):27-31

[7] FFIIS for Eurobalise, subset-036, issue 3.0, February 24, 2012. Available on the Internet, http://www.era.europa.eu/Document-Register/Documents/Set-2-Index009-SUBSET-036%20v300.pdf

[8] Saghir H, Heddebaut M, Elbahhar F, Rouvaen JM, Rivenq A, Ghys JP. Train-to-wayside wireless communication in tunnel using ultra-wideband and time reversal. Transportation Research, Part C: Emerging Technologies. 2009;**17**(1):81-97

[9] Lerosey G, de Rosny J, Tourin A, Derode A, Montaldo G, Fink M. Time reversal of electromagnetic waves and telecommunication, 17 September 2005. Radio Science. December 2005;**40**(6):1

[10] Molish AF. Ultrawideband propagation channel-theory, measurement, and modelling. IEEE Transactions on Vehicular Technology. 2005;**54**(5):1528-1545

[11] Porcino D, Hirt W. Ultra-wideband radio technology: Potential and challenges ahead. IEEE Communications Magazine. Juillet 2003;**41**(7):66-74

[12] Suwansantisuk W, Win MZ, Shepp LA. On the performance of wide-bandwidth signal acquisition in dense multipath channels. IEEE Transactions on Vehicular Technology. Sept. 2005;**54**(5):1584-1594

[13] Strohmer T, Emami M, Hansen J, Papanicolaou G, Paulraj A. Application of time-reversal with MMSE equalizer to UWB communications. Proceedings of Globecom Conference, Dallas. 2004. pp. 3123-3127

[14] Fall B, Elbahhar F, Heddebaut M, Rivenq A. Time-reversal UWB positioning beacon for railway application. International Conference on Indoor Positioning and Indoor Navigation (IPIN). 2012

[15] Bouna Fall, Fouzia Elbahhar, Marc Heddebaut, Atika Rivenq, Maria-Gabriella Di Benedetto: Assessment of the contribution of time reversal on a UWB localization system for railway applications. International Journal of Intelligent Transportation Systems Research. 2016;14(3): 139-151

[16] Fink M. Time reversal waves and super resolution. Journal of Physics. Conference series 124. 4th AIP international conference and the 1st congress of the IPIA. 2008

[17] Liu X, Wang B-Z, Xiao S, Deng J. Performance of impulse radio UWB communication based on time reversal technique. Progress in Electromagnetics Research. 2008;**79**(11):401-413

[18] Abassi-Moghadam D, TabatabaVakili D. Channel characterization of time reversal UWB communication systems. International Journal of Communication Systems. 2010;**65**(9–10): 601-614

[19] Ghavami M, Michael LV, Kohno R. Ultra Wide Band Signals and Systems in Communication Engineering. Wiley. London, 2007, 247 p

2

Transmission-Based Signaling Systems

Cesar Briso-Rodríguez,
Juan Moreno García-Loygorri and Lei Zhang

Abstract

In this chapter, we describe the principal communication systems applied to the transmission-based signaling (TBS) systems for railways. Typical examples are communication-based train control (CBTC), European Rail Traffic Management System (ERTMS), and distance to go (DTG). Moreover, to properly address some of the challenges that need to face these systems, we will provide a deep insight on propagation issues related to all the environments (urban, suburban, rural, tunnel, etc.). We will highlight all the communication-related issues and the operational as well. Finally, a detailed survey on the directions of research on all these topics is provided, in order to properly cover this interesting subject. In this research, hot topics like virtual coupling are explained as well.

Keywords: communications, engineering, railways, services, signaling, ETCS, CBTC, GSM-R, LTE, CBTC, propagation, tunnels

1. Introduction

The new signaling systems used for the control and operation of high-capacity and high-speed railways demand the use of advanced communications to guarantee their operation and safety, decrease maintenance costs, and improve the user experience. For this reason, communications in the transport environment, especially in the railway environment, have developed dramatically in the last decade.

The communications applied to railway signaling have special requirements that are very dependent on the characteristics and performances of the railway system. Consequently, it is necessary to know in detail the characteristics and performance of the transportation system and apply very strict requirements to the communication system.

This chapter describes the most important characteristics of current signaling systems and the communication systems used in combination with them.

2. Railway signaling systems

2.1. ERTMS

The European Rail Traffic Management System (ERTMS) is the most advance signaling system available nowadays. The system was developed for high-speed trains. When the train goes faster than 200 km/h, the driver is unable to see the wayside signals, so it is necessary to transmit this information to the train cabin. Moreover, this information needs to be continuous in order to have an increased safety. European Rail Traffic Management System (ERTMS) was designed to solve these two requirements and one more, the need of interoperability, letting trains to cross borders without being equipped with all the regional signaling systems (EBICAB, TBL, AWS, ASFA, LZB, and a very large etcetera). Today, there are 38 countries (most of them in Europe, but some of them not, like China, Saudi Arabia, South Korea, Taiwan, and Australia, among others), 62,000 km of track and 7500 vehicles, equipped with ERTMS [1] which is a good measure of the popularity of this system. Therefore, ERTMS leads to an increase of the interoperability, the safety, and a reduction of the costs (only one system is required).

ERTMS was specified in the European Union Agency for Railways (ERA) in a task force composed by railway manufacturers and operators (UNISIG). ERTMS is structured in three levels: level 1 is a punctual ATP with balises or loops in the track, placed 500 meters in advance of the signals. These balises/loops can provide static or variable data. Level 2 is similar to a distance-to-go (DTG) system, and the communication is now bidirectional (through the GSM-R radio), so trackside signals could be removed but not track circuits. Finally, level 3 is a "moving block" system, and both track circuits and signals could be removed. Therefore, levels 2 and 3 require GSM-R and a network of active transmitters (balises) in the track. In **Figure 1**, there is a schematic depiction of the three ERTMS levels.

Moreover, ERTMS is able to provide the driver with other information of interest, like transitions between supply phases, viaducts, and tunnels. In these areas, the train shall not stop.

2.2. CBTC

There are five grades of train automation (called GoA levels, for Grades of Automation, as it is depicted in the EN 62290 standard [2]): GoA 0, which means no automation at all; GoA 1, automatic train protection (ATP), where the train driver brakes and accelerates but under the constraints given by the system, which protects the train; GoA 2, automatic train operation (ATO), where the system regulates the speed of the train and the driver is still in the cabin doing auxiliary functions like opening and closing doors—and many others; GoA 3, driverless train operation (DTO), where no driver is needed but some staff is needed onboard the train; and, finally, GoA4, unattended train operation (UTO) mode, no staff is needed onboard.

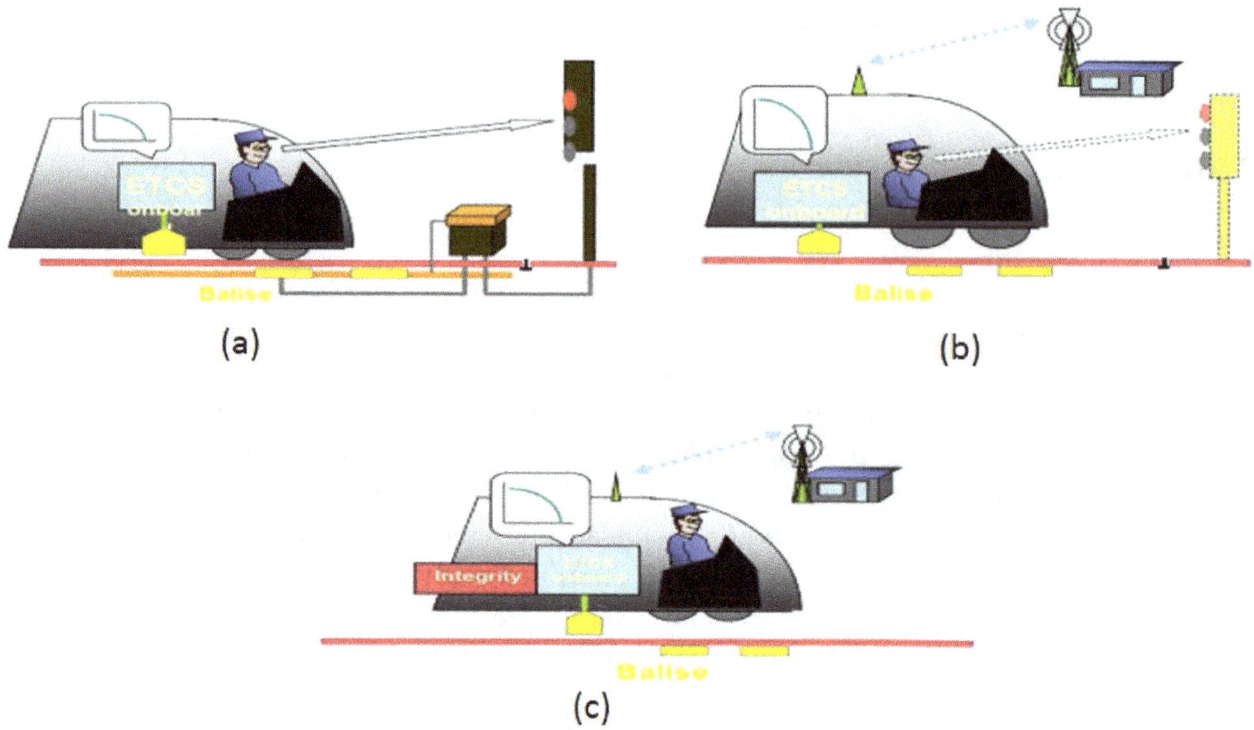

Figure 1. ERTMS levels. a) ERTMS level 1, b) ERTMS level 2, c) ERTMS level 3.

Meanwhile, metro systems have achieved all the GoA levels; mainline and high-speed trains are still in GoA 1 (or ATP). The reason behind this is that implementing higher GoA levels does not imply a significant benefit for a high-speed train operator as it is for a metro operator (headways very short, operation intensive in workforce, etc.) (**Figure 2**).

Therefore, ATO functions (speed regulation, special maneuvers, door control, etc.) are not safety related. Only ATP is, but an ATO system means 8–10% increase in the regularity of the trains compared to a human-driven ATP (plus the extra comfort for passengers due to the smoothness of speed curves).

ATP can be discrete or continuous: in the first case, the driver only receives "protection" from the system (speed monitoring and emergency brake if needed) in some certain points along the track. In the continuous mode, this is done at every single point of the track. UIC's recommendation is to implement continuous ATP systems when the maximum speed is higher than 220 km/h or the headway between trains is less than 120 seconds. Anyway, the trend is to implement this type of ATP on behalf of the safety.

Regarding the technology behind ATP functionalities, we can find for different ones: speed codes, distance to go, virtual track circuits, and moving block. In speed codes, the trackside system transmits to the onboard the maximum speed that it can achieve on a given track circuit. In distance-to-go systems, the train knows better its position in the track, which is transmitted to the next train, and this better knowledge of the position of the next train leads to a shorter distance between them. In virtual track circuits (or fixed block), the position of the train is known with a higher accuracy (less than the physical track circuit) by using odometry techniques. As it happened before, this

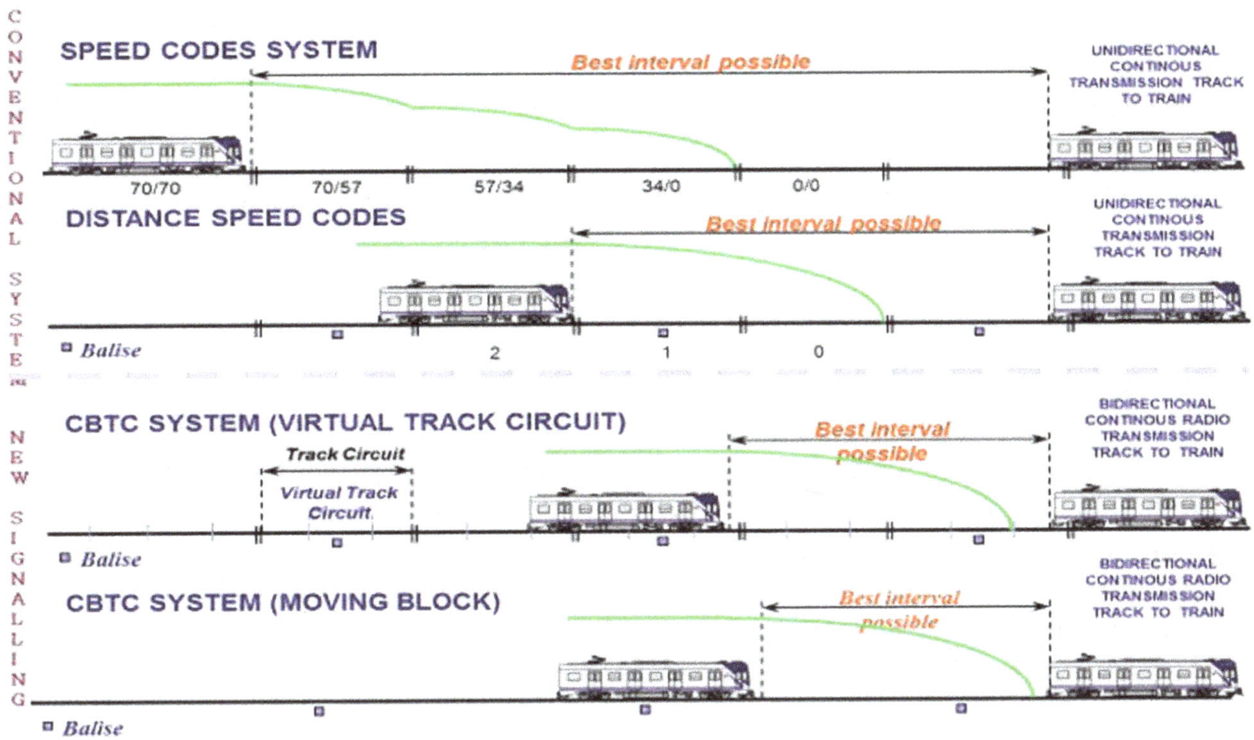

Figure 2. Comparative depiction of ATP technologies.

accuracy leads into a shorter headway. Finally, "moving block" techniques imply estimating the position of the train only using balises and odometry (no track circuits are needed), so the fragmentation due to track circuits is removed. This moving block system is the most advanced until now, and it represents 15–20% more capacity than the DTG ones [3].

DTO and UTO modes are generally implemented over communication-based train control (CBTC) systems. However, this is not a strict requirement, but an industry trend. CBTC is based on two pillars: the bidirectional communication between train and trackside equipment and the accurate positioning of the train. Train positioning is not standardized in CBTC, but it is very common to have redundant methods, like Doppler radars or tachogenerators. All CBTC vendors have their own implementation with some differences between them but with many common aspects (very high MTBFs, high positioning accuracies, etc.).

CBTC systems are able to provide headways (zero dwell, though) shorter than 60 seconds. This figure could be limited by external issues to the CBTC like delays in rail switches and other operational functions. An important remark is that there is no merit at all in having a short headway at a slow speed. The target is to have both high average speeds and short headway between trains (with no decrease in the safety).

2.3. CTCS

The Chinese Train Control System (CTCS) is a specification of train control systems of the People's Republic of China. The CTCS is based on ERTMS, and some forms are with the

European Train Control System compatibility. According to the functional requirements and equipment configuration, the application level of CTCS has divided into 0–4 levels [4] to define the equipment composition, information transmission, applicable section, track occupancy check, control mode, occlusion way, etc. in different levels.

CTCS-0: It consists of the existing track circuits, universal cab signaling, and train operation supervision system. With level 0, wayside signals are the main signals, and cab signals are the auxiliary signals. It is the most basic mode for CTCS. It is unnecessary to upgrade the wayside systems for CTCS level 0. The only way to realize the level 0 is to equip with the onboard system. CTCS level 0 is only for the trains with the speed less than 120 km/h.

CTCS-1: It consists of the existing track circuits, transponders, and ATP (automatic train protection) system onboard the train. It is used for trains with speed of 120–160 km/h. On this level, the block signals can be removed, and train operation and security are based on the onboard system, ATP, which controls the principal functions of the train: maximum speed of the track and doors opening. Transponders must be installed on the line. The requirements for track circuit in blocks and at stations are higher than that in the level 0. The control mode for ATP could be the distance to go or speed steps.

CTCS-2: It consists of digital track circuits (or analog track circuits with multi-information), transponders, and ATP system. It is used for the trains with the speed higher than 160 km/h. There is no wayside signaling in block for the level 2 anymore. The control mode for ATP is the distance to go. The digital track circuit can transmit more information than analog track circuit. ATP system can get all the necessary information for train control. With this level, fixed block mode is still applied.

CTCS-3: It consists of track circuits, transponders, and ATP with GSM-R. In the level 3, the function of the track circuit is only for train occupation and train integrity checking. Track circuits no longer transmit information concerning train operation. All the data concerning train operation information is transmitted by GSM-R. GSM-R is the core of the level. At this level, the philosophy of fixed block system is still applied.

CTCS-4: Moving block system function can be realized by this highest level. The information transmission between trains and wayside devices is made by GSM-R. GPS or transponders are used for train position. Train integrity checking is carried out by onboard system. Track circuits are only used at stations. The amount of wayside system is reduced to the minimum in order to reduce the maintenance cost of the system. Train dispatching can be made to be very flexible for the different density of train operation on the same line.

The levels 2, 3, and 4 are backward compatible with the smaller level. The CTCS-3 is functionally equivalent to the ETCS level 2. Driving in migratory space distance or absolute braking distance of CTCS-4 is also specified in ETCS level 3.

CTCS-3/ETCS and ERTMS level 2 are to be used in the People's Republic of China on the nearly 1000 km long high-speed line between Wuhan and Guangzhou. The mid-2007 awarded contract to haul equipment has a volume of 66 million euros (for installation, delivery, testing, and commissioning) and consists of the line equipment and the equipment of

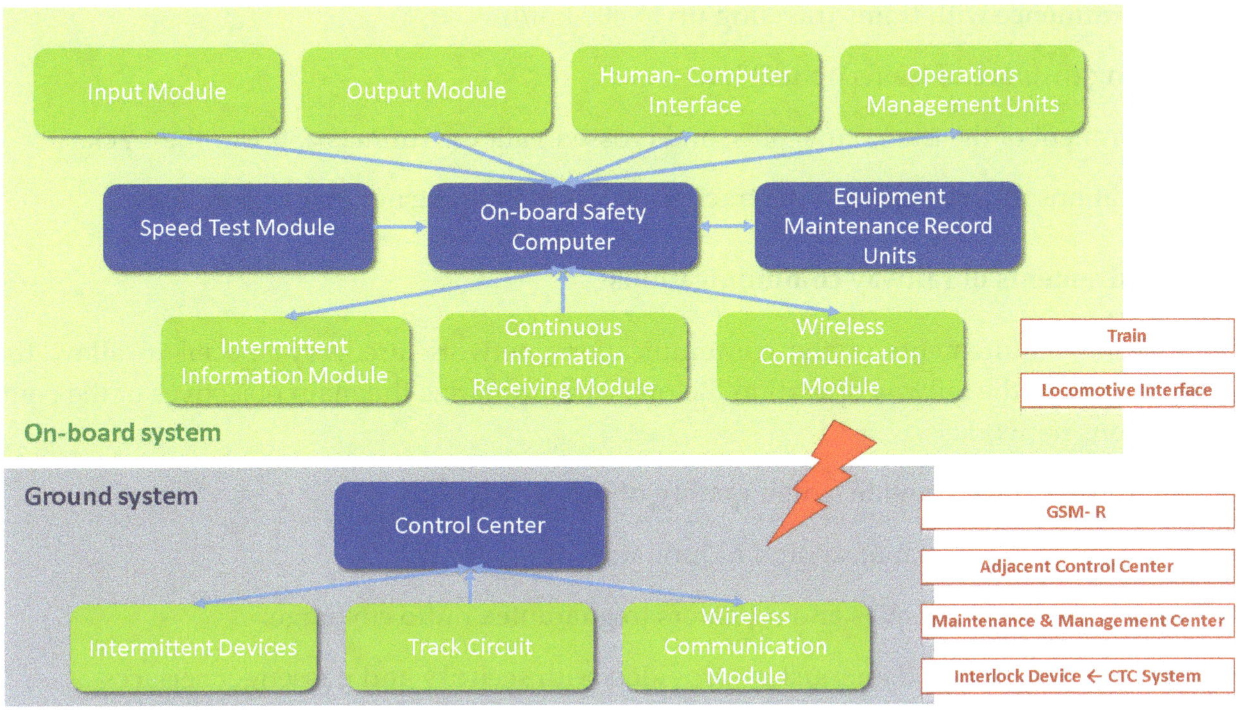

Figure 3. CTCS system architecture.

60 high-speed trains. The system has been commissioned in January 2010. The CTCS has the following features:

- **Openness**. The ETCS specification is the standard recognized by the European Union and the International Union of Railways; thus, all ETCS equipment suppliers can produce CTCS equipment in accordance with this standard.
- **Interoperability**. Since all ETCS devices are produced in accordance with uniform technical specifications, so different manufacturers of equipment can be conveniently integrated or even use directly.
- **Compatibility**. Although the vehicle equipment is different in locomotives with different levels of ETCS systems, the locomotives can operate on lines with various levels.
- **Scalability**. On the basis of the original equipment in low-level CTCS system, it can be easily upgraded to a higher level by adding some new equipment (modules); the original train control equipment can be continually used in high-level systems (**Figure 3**).

3. Communications in railway environment

The railway environment is very complex and represents a major challenge for communication systems. For this reason, to deploy a communication system in this environment, it is necessary to understand their special characteristics. The main features of the railway environment are summarized below:

- Speed influence with trains traveling up to 400 km/h.
- Breaking distance of trains up to 3 km.
- Environment of the track: cuts, hills, and crossed valleys. Tunnels >30% of the track.
- Base stations must be close to the track to access the infrastructure of the railway.

3.1. Requirements of railway communications

The general requirements of the communications network are very special to allow the improvement of the railway operation. Below, we summarize the main objectives of the communications network:

- The network must provide a high quality of service (QoS).
- The design must guarantee a high redundancy of coverage.
- Trains must have antennas and repeaters to guarantee radio coverage.
- Signaling and control data must be provided with security and passenger services.
- There are important limitations in the frequency bands that can be used.

Also, communication networks for railway applications must provide special services to enable railway operations and ensure safety. The main services are summarized below:

- Group broadcast call: multipoint link, emergency call.
- Fast call setup ⇒setting calls <2 seconds (Eirene).
- The use of a special or proprietary frequency band.
- Improved equalization ⇒ mobile traveling at 350 km/h.
- Location-based addressing—training calls originated on trains depending on location.
- Functional addressing ⇒ numbers for each function of the train.

The principal specifications of the network based on the European Signaling System are summarized in **Table 1** [5]. These speculations are used to validate the communications network in

ETCS QoS parameter	Requirement
Connection establishment	<8.5 s (95%), ≤10 s (100%)
Transfer delay (TD)	≤0.5 s (99%)
Network registration delay (NRD)	≤30 s (95%), ≤35 s (99%), ≤40 s (100%)
Connection loss rate (CLR)	<10^{-2}/h
Transmission interference (TI)	<10^{-2}

Table 1. Main requirements of railway communication system.

a railway environment. The compliance of these specifications is verified by measuring along the track and averaging every 100 m. This guarantees the quality of service and reliability.

3.2. Network architecture

In order to guarantee the quality and reliability of the communications in the railway environment, it is necessary to design the networks to ensure an excellent coverage in percentages above 99% of the time and locations:

- Dual-layer or single-layer coverage with strong overlap.
- Two totally independent layers: 2 MSC, BSC, TRAU.
- Two totally independent redundant systems for the fixed network.
- Hot standby: change from layer A to layer B without loss of call.

One of the main requirements of the GSM-R network is to achieve a high quality of service. For this, two different strategies can be used:

- Two complete networks overlapped.
- A single network with a strong overlap of the radio coverage of base stations (**Figure 4**).

In the first configuration, two overlapped networks are used. This allows that if one or several base stations of one layer fail, an automatic roaming occurs to the other layer, thus ensuring communications. In the second design, a network with a double number of base stations is used. This allows a strong overlap of the radio coverage, so that if one cell fails, neighbor cells compensate its coverage automatically. In both cases, a high redundancy is achieved that guarantees the operation of the system.

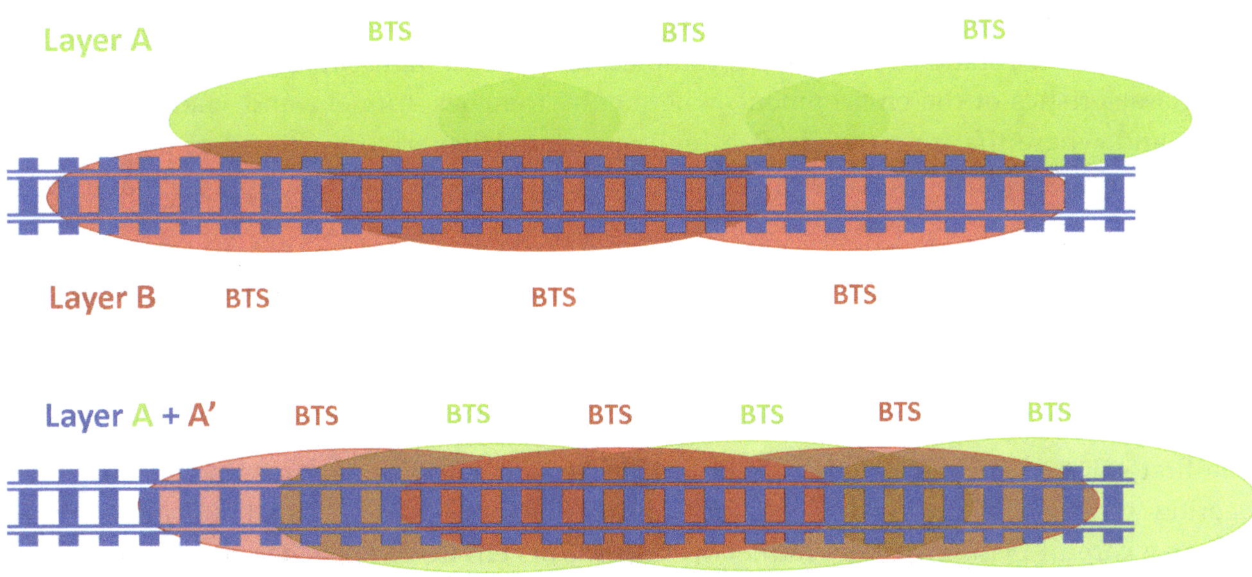

Figure 4. Radio coverage for high QoS and redundancy.

3.3. Communication problems in railway environments

As already mentioned, railway environment is very complex, and it is necessary to provide communications in a variety of environments and conditions that make it very difficult to plan and deploy a radio communication system with the quality of services demanded by a signaling system. The different communication systems used on railway environment have reported several problems on the different test trials made. The principal problems reported are summarized as follows:

- Failure of handovers: at high speed (250–350 km/h), some handovers can be lost. These failures depend also on the direction of the train. The main cause of the loss of handover is the little overlap of the radio coverage of neighbor cells. For this reason, the train loses coverage too fast, without completing the handover that requires 6 seconds.
- Ping-pong: this phenomenon is due to the realization of several handovers with the neighboring cell. It happens because the overlap with the neighboring cell is not uniform, so the coverage of the cell is better in some points and worst in others. For this reason two or more handovers are made when only one should happen. This significantly deteriorates the QoS and causes link losses.
- Loss of data during the handover: it is due to the cut of the link when the handover takes place between two neighbor cells. Depending on the type of communication system can have a length from some millisecond to 500–700 milliseconds.
- Bit error rate (BER) degradation with speed: communication systems employ channel equalizers to mitigate propagation problems and reduce transmission errors. However, these equalizers require the channel to remain stable for a short period of time. For this reason at high speeds, the channel varies too fast, and transmission errors occur, which increases the BER.

Another important aspect of railway communications is the problem of the railway environment. In this environment, providing a radio coverage with high QoS is complicated due to the characteristics of the environment, where cuts, trenches, tunnels, and viaducts are frequent and where the train travels at a high speed. Particularly, the study of communications in tunnels, which can currently account for up to 50% of the route of a railway line, is important.

Communication in tunnels is a major challenge for radio communication systems; so in this environment, it is necessary to use solutions other than those used in open areas. Basically, two different techniques can be used to provide coverage in tunnels: antennas and leaky feeder.

The first solution is based on the use of antennas. This solution requires much more engineering work, but it has less deployment and maintenance cost. First of all, it is necessary to ensure a careful planning of the radio coverage in this environment. Then, it is necessary to model the propagation in detail obtaining a propagation model adjusted to the characteristics of each tunnel and the communication system used. Then, it is necessary to validate the

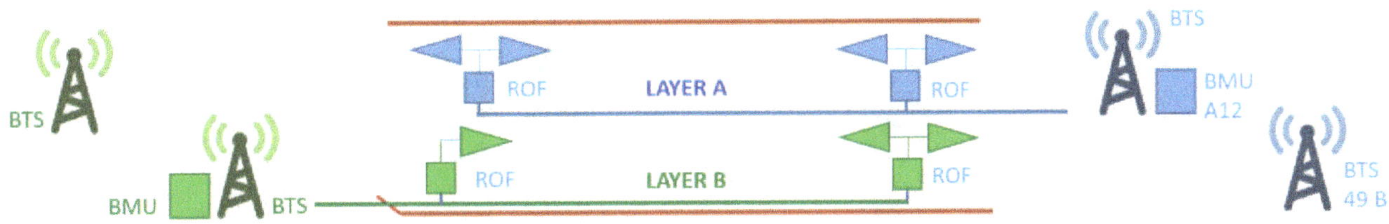

Figure 5. Radio coverage of a tunnel with repeaters and antennas.

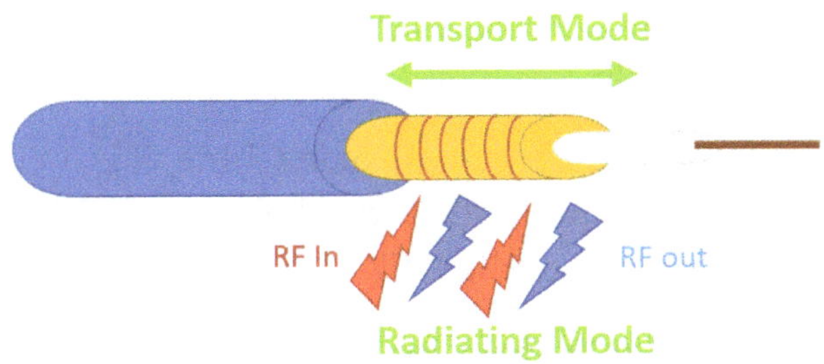

Figure 6. Leaky feeder.

approach with measurements on the target tunnel, and finally a solution is custom designed for each tunnel. In **Figure 6**, a solution for radio coverage of a railway tunnel is shown. In this case, the solution is based on the use of repeaters with technology radio over fiber [6] connected with an optical fiber. Each repeater retransmits the signal of the nearest base stations inside the tunnel, so the radio coverage is extended along the tunnel. In the case of **Figure 6**, two repeaters for each network layer are used to provide redundancy inside the tunnel. The handover is made when the train goes outside the tunnel. This solution requires the installation of a repeater every 1–2 km of the tunnel (**Figure 5**).

The second possible solution for radio coverage in tunnels is the use of leaky feeders [7]. In this case, a low loss coaxial with slots on the shieling (see **Figure 6**), is deployed along the tunnel wall. The antennas of previous case can be replaced by the coaxial. The result is a much stable radio coverage with a power level stable at a distance of 2 m from the cable. The disadvantages are the high cost of this solution that also needs a shorter distance between repeaters.

4. Wireless standard for railway

4.1. GSM-R

GSM-R system description is essentially the same system as the GSM but with railway-specific functionalities. It uses a specific frequency band around 800/900 MHz, as illustrated in **Figure 7**. In addition, the frequency bands 873–876 MHz (uplink) and 918–921 MHz (downlink) are used as extension bands for GSM-R on a national basis, under the name extended GSM-R (E-GSM-R). GSM-R is typically implemented using dedicated base stations (BSs) close to the rail track.

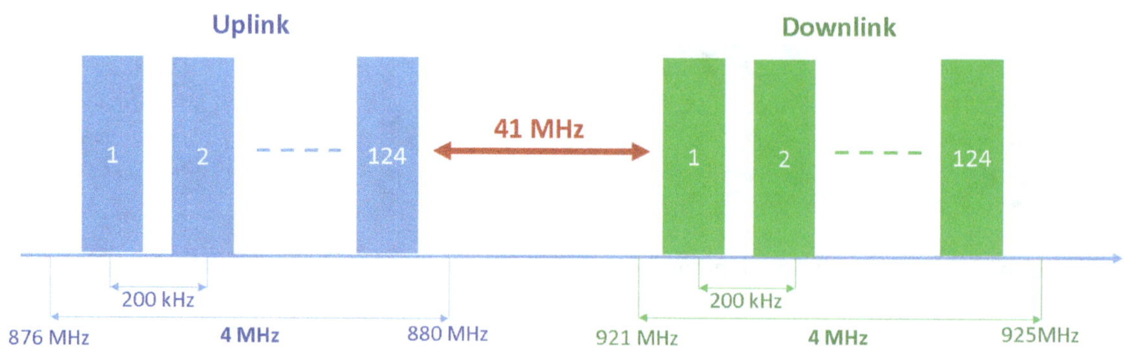

Figure 7. GSM-R frequency band and channels.

The distance between two neighbor BSs is 7–15 km; in China, it is 3–5 km because redundancy coverage is used to ensure higher availability and reliability. GSM-R has to fulfill tight availability and performance requirements of the HSR radio services.

The GSM-R network serves as a data carrier for the European Train Control System (ETCS), which is the signaling system used for railway control. The ETCS has three levels of operation and uses the GSM-R radio network to send and receive information from trains. On the first level, ETCS-1, the GSM-R is used only for voice communications. On the other two levels, ETCS-2 and ETCS-3, the GSM-R system is used mainly for data transmissions. The GSM-R is very relevant to ETCS-2 and ETCS-3, where the train travels at a speed up to 350 km/h, and it is thus necessary to guarantee a continuous supervision of train position and speed. When the call is lost, the train has to automatically reduce the speed to 300 km/h (ETCS-1) or lower.

The most typical HSR-specific services offered by GSM-R are as follows [8]:

- Voice group call service (VGCS): VGCS conducts group calls between trains and BSs or conducts group calls between trackside workers, station staff, and similar groups.
- Voice broadcast service (VBS): The BS broadcasts messages to certain groups of trains, or trains broadcast messages to BSs and other trains in a defined area. Compared to VGCS, only the initiator of the call can speak in VBS, and the others who join the call can only be listeners. VBS is mainly used to broadcast recorded messages or to make announcements in the operation of HSR.
- Enhanced multilevel precedence and preemption (eMLPP): eMLPP defines the user's priority and is used to achieve high performance for emergency group calls.
- Shunting mode: Shunting mode provides an effective means of communication to a group of personnel who are involved with a shunting operation, which regulates and controls user access to shunting communications (a link assurance signal used to give reassurance to the train driver).
- Functional addressing: A train can be addressed by a number identifying the function for which it is being used rather than a more permanent subscriber number.
- Location-dependent addressing: Calls from a train to certain functions can be addressed based on the location of the train as the train moves through different areas of BSs (**Figure 8**).

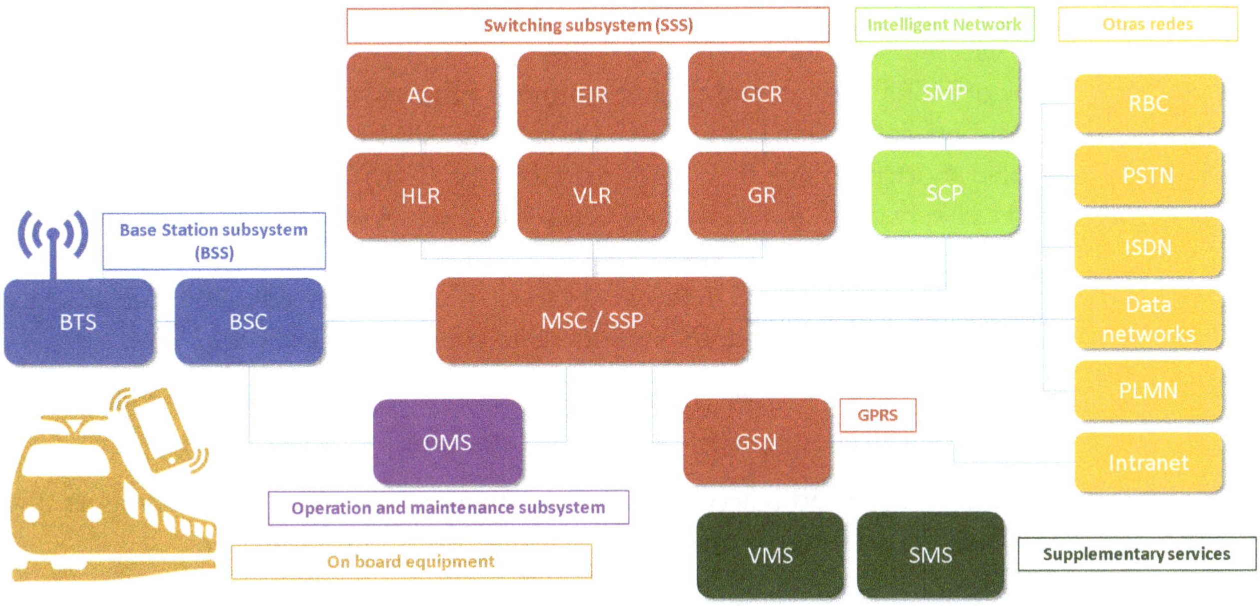

Figure 8. GSM-R network.

4.1.1. GSM-R limitations

Although the popularity of GSM-R is still growing, increasing interference from public networks is hampering the use of GSM-R, while the assigned radio frequencies limit its capacity. Several limitations are summarized in the following:

- **Interference**: The interference between GSM-R and other public networks increases because both railway and public operators want to have good coverage along the rail tracks. Instead of cooperating in network planning, railway and public operators fight for the coverage. The interference could result in severe impairment of voice and data communications as well as network loss over several hundred meters of track. Theoretically, such interference can be avoided if public operators do not use frequency bands adjacent to those of GSM-R for the areas close to rail tracks; however, this is not well implemented in practice. In the future, interference may increase owing to the growth of GSM-R network deployment and the potential growth of public networks.

- **Capacity**: The 4 MHz bandwidth of GSM-R can support 19 channels of 200 KHz width. This is sufficient for voice communication, as voice calls are limited in time and do not occupy resources continuously. However, the current capacity turns out to be insufficient for the next-generation railway system, where each train needs to establish a continuous data connection with a radio block center (RBC) and each RBC connection needs to constantly occupy one time slot. The radio capacity can be increased by using more spectrum resources.

- **Capability**: As a narrowband system, GSM-R cannot provide advanced services and adapt to new requirements. The maximum transmission rate of GSM-R per connection is 9.6 kb/s, which is sufficient only for applications with low demands; message delay is in the range of 400 ms, which is too high to support any real-time application and emergency communication [9].

4.1.2. Future of HSR communications

The future services of HSRs such as real-time monitoring require a wideband system to have larger data rate and short delay. Due to the above limitations, GSM-R must eventually evolve to eliminate the revealed shortcomings [10]. Long-term evolution (LTE)-R, which could be based on the LTE standard, is a likely candidate to replace GSM-R in the future for the following reasons:

1. LTE has many advantages over GSM in terms of capacity and capabilities. As a fully packet-switched–based network, LTE is better suited for data communications.
2. LTE offers a more efficient network architecture and thus has a reduced packet delay, which is one of the crucial requirements for providing ETCS messages.
3. LTE has a high-throughput radio access, as it consists of a number of improvements that increase spectral efficiency, such as advanced multiplexing and modulation.

LTE is also a well-established and off-the-shelf system and provides standardized interworking mechanisms with GSM [11].

4.2. TETRA

TETRA is an ETSI standard [12] for public safety, very popular in police, civil defense, border control, fireman, etc. It allows point-to-multipoint and point-to-point (direct) communications, of both voice and data. It is a very robust system, popular in urban railways but very limited because it is a narrowband system (up to some kbps). This technology usually works in the 380–470 MHz band, which needs a smaller number of base stations than GSM-R (which works in the 900 MHz band) for a given distance. TETRA has four 25 KHz channels, which is not enough for high data rates. The industry trend is to replace TETRA systems with LTE.

4.3. IEEE 802.11

It is almost impossible to gather all the complexity of the IEEE 802.11 standards [13] in so little space. IEEE 802.11 is a worldwide famous standard (its popular name is "Wi-Fi") which constitutes one of the most successful communication systems ever. It specifies both the physical and the MAC layer and has had many amendments since its beginnings. A summary of the most important features of the main standards is presented in **Table 2**.

Characteristic	IEEE 802.11b	IEEE 802.11g	IEEE 802.11n	IEEE 802.11ac
Frequency band	2.4 GHz	2.4 GHz	2.4 and 5 GHz	5 GHz
Channel width	20 MHz	20 MHz	20 and 40 MHz (only in 5 GHz)	20, 40, 80, and 160 MHz
Transmission tech.	SISO	SISO	MIMO	MIMO
Max. rate (Mbps)	11	54	433	1300

Table 2. Main characteristics of the IEEE 802.11 family of standards.

IEEE 802.11 technologies are widely used in railways for two main reasons: they work in unlicensed bands (ISM, ETSI, etc.) and its low cost. The two main applications for them are train-to-ground communications (in workgroup bridge mode) and access to the Internet for passengers (client: access point (AP) mode). It is worth mentioning that the workgroup bridge mode is not standardized, so it has some proprietary aspects that could break the interoperability that is assumed in the usual client: AP mode.

4.4. 4G/5G

In this section, we will cover the feasibility of the fourth and the fifth generation of mobile communications to provide railway signaling services. LTE (long-term evolution), the most relevant 4G technology, is standardized by the 3GPP. LTE has a flat architecture, to suppress bottlenecks and also to achieve low end-to-end delays. LTE vendors like Huawei, Nokia, or ZTE have a lot of interest in railways, and some years ago, the UIC identified LTE as a key technology for railway communications [14]. Explaining LTE in some detail is out of the scope of this chapter, but there are very good references in the literature worth reading [15, 16]. The most remarkable features of LTE are that it is a full-IP technology; is very flexible that can work either in TDD or FDD mode; and supports carrier aggregation and several MIMO configurations. Its general architecture is shown in **Figure 9**, with an access stratum, composed by eNBs, and the core. More details can be found in [16].

There are still many challenges to have a successful LTE deployment for signaling. The first one is the spectrum, because almost everywhere there is no frequency band allocated for this purpose. Almost every single LTE band is licensed, so the railway operator would need to reach an agreement with a mobile operator in order to provide this service with the needed requirements for signaling services. Among the other challenges, we can find the cybersecurity (almost every single signaling network is owned by a railway operator) and the cost (generally speaking, LTE is more expensive than other technologies that do not need a core, but this is a controversial statement).

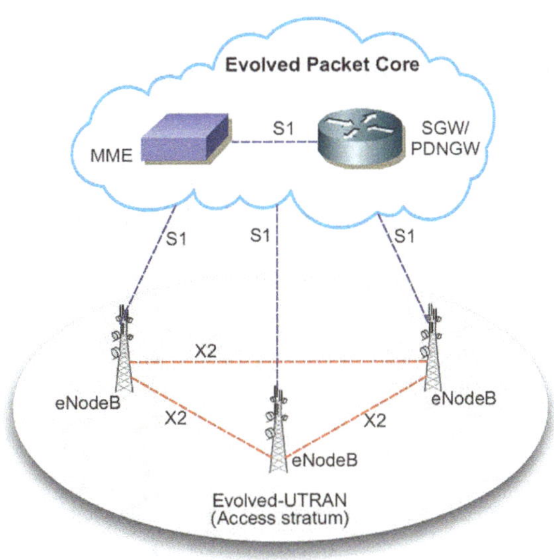

Figure 9. LTE architecture.

Regarding 5G mobile communications, it is still a technology not in operation but being massively researched. It is very promising not only for railways but also for the general industry and civilian uses. For railways, it is expected that 5G will be an enabler for the next generation of services, like "virtual coupling" (which needs low latencies and high reliability), "remote driving" (both high capacity and low latencies needed), and a large etcetera.

5. Wireless channel in railway

On conventional railway communication, the high penetration losses caused by the train's body, the harsh electromagnetic environment, and diverse scenarios have brought many difficulties. In recent years, the rapid development of HSTs makes demands on broadband transmission, high capacity, and reliability services regardless of the locations and speeds. Thus, modern wireless communication systems for railway have to overcome additional challenges, such as fast handover, fast travel through diverse scenarios, and large Doppler spreads, resulting from the high speed of the train (>250 km/h) [17].

In this section, the basic knowledge of the large-scale fading and small-scale fading of radio propagation is presented. Also, the channel characterization and modeling for railway are discussed to provide a brief overview of the research status for wireless channel in various railway scenarios.

5.1. Radio propagation mechanism

The fading in mobile communication can be classified into two main groups: large-scale fading and small-scale fading. Literally, the large-scale fading denotes a large distance (hundreds of wavelengths) that the radio signal traveled. Generally, the slow dissipation of energy due to the separation of transmitter and receiver within such a large distance is defined as path loss. In the meantime, the penetration or diffraction around large objects in the actual propagation channel results in the fluctuations on the local mean power, which are the so-called large-scale fading (or "shadowing"). Small-scale fading is used to describe the self-interference of the arrived signals from different paths with different amplitudes, delays, and phases at the receiver over a short period (in the order of 10 ns depending on mobility) or a travel distance on the same order of the wavelength. **Figure 10** illustrates the large-scale fading that superimposed with the small-scale fading.

When radio signal travels through the wireless channel in the railway environment, the presence of trees, people, buildings, mountains, and other obstructions results in the signal

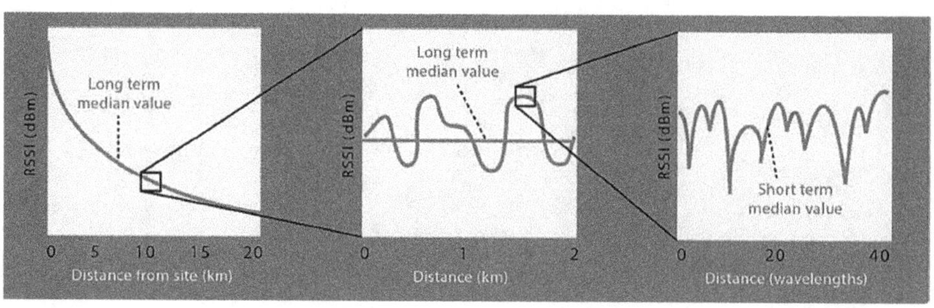

Figure 10. Path loss, shadow fading, and small-scale fading.

undergoing many kinds of propagation effects that result the radio signal reaches at the receiving antenna by two or more paths. This multipath denotes that the radio signal is transmitted from the transmitter and then "interacted" (reflected/diffracted/scattered) with the objects (also known as "scatterers") presented in the channel. The multiple copies of the transmitted signal travel from several different paths and then arrive and combine at the receiver. This combination can be either constructive or destructive that can cause random fluctuations in the received signal's amplitude, phase, and angle of arrival, which is termed as multipath fading.

5.2. Channel characterization and modeling for railway

A channel model is known as an abstract and simplified approach to mathematically demonstrate the main characters of an actual channel and evaluates the influences on the performance of a specific wireless technology in this channel. Fundamentally, the modeling approaches fall into two categories: physical models and analytical models. The physical models are established on the basis of electromagnetic wave propagation and independent of antenna configurations, such as antenna pattern, number, etc. It models the bidirection propagation between the transmitter and receiver based on the measurement or simulation in a specific scenario, which means that it changes as the locations of transmitter and receiver changes, whereas the analytical models mathematically/analytically reproduce the statistical properties of the MIMO matrix in the corresponding domain, which is limited by the computation complexity. The physical channel models are widely adopted and generally classified into two types: (1) **Deterministic modeling** characterizes the physical propagation in a completely deterministic way, i.e., calculating the received signal from knowledge of geometry, the electrical properties of the medium of propagation, cross section of objects, and antenna radiation pattern. Therefore, the deterministic models are approximately exact. However, the deterministic approach, such as ray tracing, is subject to the computation complexity and complication of environment reconstruction. (2) **Stochastic modeling** is void of exact geometrical assumption and provides the statistical manner and conditional dependencies between different channel parameters with less computation resources. This kind of channel model is extracted from the large amounts of measurement data, which is practically useful to present the general characteristic of a typical scenario. The propagation characterization in the specific scenarios is modeled by the statistical analysis of the channel impulse response (CIR), extracting the key parameters, such as path loss, delay spread, angular spread, etc., to parametrize the channel (**Figure 11**).

The wireless channel characteristics in railway are of importance for train control and passenger services. The deep and comprehensive knowledge of the propagation channel is the premise of the communication system design. A convincible solution to support wireless communication system design for railway requires the characterization of four different channels, as shown in **Figure 10**: BTS-to-train (B2T)/train-to-ground (T2G), carriage-to-carriage (C2C), train-to-train (T2 T), intra-carriage (InC), and optionally satellite communication train-to-satellite (T2S). In the last years, research on communication systems for railway has been focused specially on the T2G communications in different environments, as presented in [18, 19] with the complete surveys on results and challenges of high-speed trains (HST) and metro trains in tunnels, respectively.

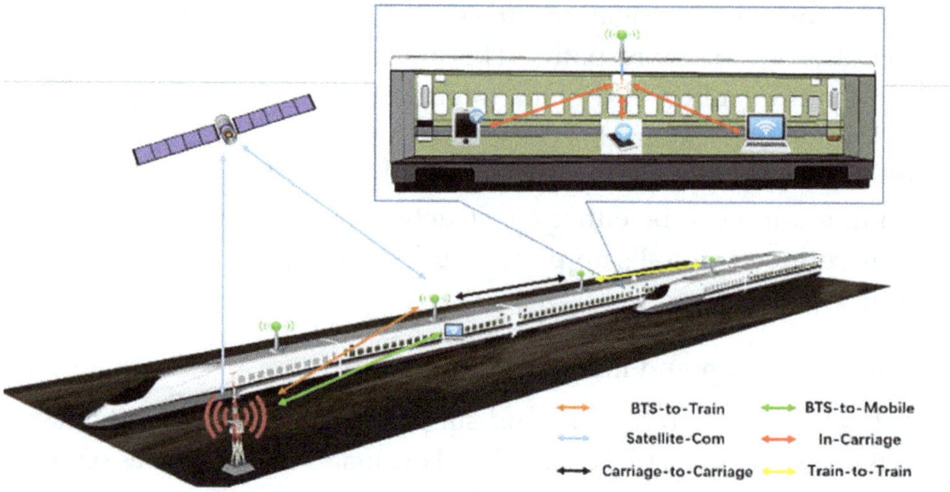

Figure 11. Wireless links in railway.

Scenario	Path loss	Shadowing	K-factor	Delay	Doppler	AoA/AoD	EoA/EoD
Open space	M & S	M & S	M & S	M & S	M & S	M & S	—
Viaduct	M & S	M & S	M	M	M	M	—
Cutting	M & S	M & S	M	M	M	—	—
Metro	M	M	M	M	—	—	—
Intra-carriage	M	M	M	M	—	—	—

Table 3. The research status of the channel characterization and modeling for railway.

In recent years, the channel characterization and modeling have become the focus of a lot of studies based on the measurement campaigns. Subject to the robustness, scalability, hardware redundancy, and traceability of the particular hardware, measurement-based approaches cannot keep the spatial consistency which is crucial in time-varying channel modeling in railway. Thus, some deterministic channel models have been proposed based on the ray tracing. Here, we briefly review and classify the measurement (**M**)-/simulation (**S**)-based research status for railway communications according to the scenarios, approaches, and key channel parameters, including path loss, time delay, and so on, as shown in **Table 3** [17, 20].

6. Future trends in signaling systems

It is always hard to predict the future, even the more immediate one. Signaling, like the entire railway sector, goes in the same direction: more efficiency, more integration between systems, open standards (where it is possible), and more safety and security. When we described ERTMS, the lack of implementation of its level 3 was noteworthy. The reasons for this are many, but perhaps the most relevant is that it implies a translation of costs from infrastructure managers to railway operators (less wayside equipment needed, but some functions to be done by trains, like guaranteeing the train integrity). However, in the last

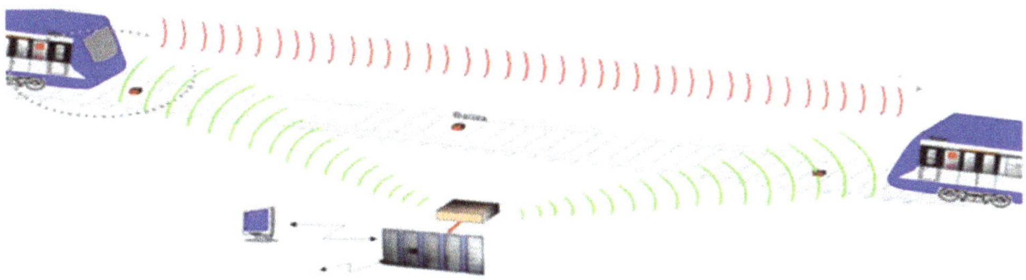

Figure 12. Schematic of the V2V-CBTC with both a direct link between trains and a train-to-infrastructure link.

years, some tests [21] have been carried out to check the feasibility of the level 3 of ERTMS, but not following a satisfactory approach (some parts were missing in these proofs of concept). Moreover, this is one of the main projects within Shift2Rail IP2 [22].

Another trend for the future of ERTMS is its "low-cost versions." Under this denomination we can find some initiatives that are focused on the implementation of a standardized and interoperable ERTMS but with some limitations to lower its CAPEX and OPEX. The most remarkable initiative in this direction is the "regional ERTMS" [23] where the line is divided into "dark zones"; only one train is allowed into each one of them. More relevant, there is no need of track circuits, and GSM-R is needed only in a punctual way.

When we explained CBTC, we saw that it can work in two modes: virtual track circuit and "moving block." The latter is very similar to ERTMS-L3; so in metro lines, this paradigm has been achieved some years ago. The next step in CBTC systems is to have a direct link between trains, in order to decrease the communication delay and the headway between them as well. Some signaling companies are working in this idea, but, as far as we know, only Alstom has shown some interest on it. In **Figure 4**, the basic concept under this "vehicular-to-vehicular CBTC" is shown. It would not remove the existing link between trains and wayside, but to be complemented by a direct link (V2 V) between trains when they are closer enough (**Figure 12**).

Finally, the sublimation of the signaling models is the so-called virtual coupling [23], where a large set of trains not coupled together could act as if they were. This implies sharing the same braking and power information for the entire "virtual train" that could be a large number of trains. To achieve this futuristic model, we need an ultrareliable communication link, with a very low latency. This line of research was raised 20 years ago but now is being researched in the EU-funded Shift2Rail IP2 initiative [22], with the aim of having a demonstrator with real trains in 2022.

Author details

Cesar Briso-Rodríguez[1]*, Juan Moreno García-Loygorri[1] and Lei Zhang[2]

*Address all correspondence to: cesar.briso@upm.es

1 Universidad Politecnica de Madrid, Spain

2 Key Lab of Wireless Sensor Network and Communication, Shanghai Research Center for Wireless Communications, China

References

[1] http://www.railway-technology.com/projects/european-rail-traffic-managementsystem-ertms

[2] EN 62290-1:2014 Railway applications. Urban Guided Transport Management and Command/Control Systems. System Principles and Fundamental Concepts

[3] González FJ. Señalización ferroviaria: del guardagujas a la operación sin conductor; Fundación Dialnet. 2008

[4] WIT press. Advanced Train Control Systems. 1st ed. WIT press; 2010

[5] European Railway Agency. ERTMS GSM-R QoS Test Specification [Internet]. July 2016. Available from: http://www.era.europa.eu/Document-Register/Pages/O_2475.aspx

[6] Tetsuya K, Atsushi K, Pham TD, Naokatsu Y. Seamless access networks using radio-over-Fiber. In: IEEE, editor. 2016 IEEE Sixth International Conference on Communications and Electronics (ICCE); India. 2016

[7] Motley AJ, Davis QV. Reliable high-speed data transmission from a mobile vehicle via a leaky feeder. Radio and electronic. Engineer. 1980;**50**(7):337-344

[8] Ruisi H, et al. High-speed railway communications: From GSM-R to LTE-R. IEEE Vehicular Technology Magazine. 2016;**11**(3):49-58

[9] NOKIA Networks. Railways. [Internet]; 2017. Available from: https://networks.nokia.com/railways

[10] Jaime C, et al. Long term evolution in high speed railway environments: Feasibility and challenges. Bell Labs Technical Journal. 2013;**18**(2):237-253

[11] Aleksander S, Jose. LTE for railways: Impact on performance of ETCS railway 999. IEEE Vehicular Technology Magazine. 2014;**9**(2):69-77

[12] Terrestrial Trunked Radio (TETRA). Voice plus Data (V+D); Designers' Guide; Part 1: Overview, Technical Description and Radio Aspects http://www.etsi.org/deliver/etsi_etr/300_399/30001/01_60/etr_30001e01p.pdf

[13] IEEE 802.11 standards. http://standards.ieee.org/about/get/802/802.11.html

[14] LTE/SAE -The Future Railway Mobile Radio System: Long-Term Visions on Railway Mobile Radio Technologies, UIC (Unión International Communications), V 0.4 Draft 14.09; 2009.

[15] [15GPP] Specifications [Online]: http://www.3gpp.org/dynareport/36-series.htm

[16] Dahlman E, Parkvall S, Skold J. 4G: LTE/LTE-Advanced for Mobile Broadband. 2nd ed; Elsevier. 2013

[17] Wang CX, et al. Channel measurements and models for high-speed train communication systems: A survey. IEEE Communications Surveys & Tutorials. 2016;**8**(2):974-987

[18] Ai B, et al. Challenges toward wireless Communications for High-Speed Railway. IEEE Trans. Intelli. Transp. Sys. Oct 2014;**15**(5):2143-2158

[19] Hrovat A, et al. A survey of radio propagation Modeling for tunnels. IEEE Communications Surveys & Tutorials. 2014;**16**(2):658-669

[20] Chen B, et al. Channel characteristics in high-speed railway: A survey of channel propagation properties. IEEE Vehicular Technology Magazine. 2015;**10**(2):67-78

[21] Furness N, van Houten H, Arenas L, Bartholomeus M. ERTMS level 3: The game-changer. IRSE News. April 2017;**232**:2-9

[22] Innovation Programme 2 – Shift2Rail. https://shift2rail.org/research-development/ip2

[23] ERTMS Regional – Status and Future Enhancements, UIC ERTMS World Congress Istanbul, Turkey 1st–3rd. April 2014. http://www.ertmsconference2014.com/assets/SESSION-PRESENTATIONS/S2/ERTMS-Regional-Status-andfuture-enhancements-2014-01-27-ed01.pdf.

3

Application of Multicriteria Decision-Making Methods in Railway Engineering: A Case Study of Train Control Information Systems (TCIS)

Boban Djordjević and Krmac Evelin

Abstract

In order to improve its position in the transport market railway, as a complex system, it has to fulfill a number of objectives such as increased capacity and asset utilization, improved reliability and safety, higher customer service levels, better energy efficiency and fewer emissions, along with increased economic viability and profits. Some of these objectives call for the implementation of maximum values, while some of them require minimum values. Additionally, some can be expressed quantitatively, while some, for example, customer service, can be described qualitatively through a descriptive scale of points. The application of MCDM in railway engineering can play a significant role. Therefore, the major objective of this chapter is the review of the application of MCDM methods in railway engineering. As one of the means in achieving the objectives of railways and above all the utilization of capacity are Train Control Information Systems (TCIS). Based on that, the aim of this chapter is the evaluation of the efficiency of TCIS in the improvement of railway capacity utilization through defined technical-technological indicators. The non-radial Data Envelopment Analysis (DEA) model for the evaluation of TCIS efficiency in improvement of utilization of railway capacity using the selected indicators is proposed. The proposed non-radial DEA model for TCIS efficiency evaluation in using railway capacity could be applied to an overall network or for separate parts of railway lines.

Keywords: evaluation, efficiency, railway, capacity, train control information system, non-radial DEA model

1. Introduction

1.1. Background

Transport activity in the European Union (EU) is expected to continue growing but with slower trends than in the past [1]. From 1995 to 2012, freight transport in the EU28 countries showed an overall growth of 22.8%, while passenger activity was slower with 19% [2]. According to the [1], the increment of freight transport activity is expected to be 35% from 2010 to 2030 and 58% from 2010 to 2050, while passenger activity growth will be slightly lower at 22% until 2030 and 40% until 2050. Railway as a backbone of the EU transport system has the potential to play a key role in addressing rising traffic demand, congestion, fuel security and decarbonisation, but many European rail markets are still facing stagnation or decline [1].

The demand for railway transportation is steadily increasing around the world [3]. Regarding the EU, railway transport activity is expected to grow in terms of passenger transport by 39% between 2010 and 2030 and 76% from 2010 to 2050, as well as freight transport by 47% by 2030 and 84% during 2010–2050. Therefore, railway share of the transport market related to passenger transport will be increased to 1 percentage point by 2030 and an additional percentage point by 2050. Similar increases in modal share for passenger rail are expected for freight transport [1].

However, because the many railways use capacity close to maximum, it is necessary to take adequate actions in order to meet the new demand. Accordingly, possible actions could be building a new railway infrastructure, upgrading existing infrastructure or using the existing infrastructure more efficiently [3]. Consequently, the main challenge of many railways around the world is limited availability of capacity for all trains of their infrastructure related to topological configuration; although in some cases the capacity of infrastructure was not changed despite doubling, tripling or quadrupling tracks.

Railway capacity may be defined in different ways [3]. Indeed, in the railway market and in the literature, there are numerous definitions and meanings of railway capacity [4, 5]. Railway capacity is seen as a simple but rather inaccessible concept. Although unique, a true definition of capacity is impossible, because capacity depends on the way it is utilized. An overview of some important definitions of railway capacity was presented in [6]. One of the most appropriate definitions of railway capacity for this chapter was declaimed by [4]. They stated that railway "capacity is a measure of the ability to move a specific amount of traffic over a defined rail line with a given set of resources under a specific service plan."

In order to tackle the railway capacity challenge, since the building of new railway lines is extremely costly and time-consuming, efficient utilization of railway infrastructure is essential. Therefore, there is a need for better planning and efficient management of measuring used capacity, as well as better enhancement measures [6]. Analysis and evaluation of railway capacity is a multifaceted and complex task because it involves several complex systems, such as railway infrastructure, rolling stock, timetable and human behavior, and some other influencing factors such as the number of tracks connecting the stations, station track layout, the signaling system, train performance, speed difference between train services, market demand, reliability and delay acceptance of railway customers.

As a multidisciplinary area different factors impinge, such as timetable, signaling, nodal capacity constraints, rolling stock, infrastructure, external factors and governance impact on railway capacity utilization. Signaling as a kind of traffic management systems (TMS) and one of the classes of train control information systems (TCIS) ensures the safe running of trains within the infrastructure. With advanced signaling systems, such as European Railway Traffic Management System (ERTMS), besides the improvement level of safety, reduction of headway and blocking time of infrastructure may be achieved [6, 7].

According to [7], TCIS represent a complex system composed of a large number of various kinds of components (mechanical, electrical, computer, etc.) with different types of interactions (local, simultaneous, etc.), which are interconnected and operate in synergy with each other. In their chapter, TCIS are grouped in four classes where each class includes subsystems, that is, Interlocking Systems (IXL) including subsystems such as train detection equipment (track circuits or axle counters), switching points and level crossings; Traffic Management Systems (TMS) that imply ERTMS, Positive Train Control System (PTCS), Chinese Train Control System (CTCS) and Railway Signaling System (RSS); Automatic Train Control (ATC) systems including Automatic Train Protection (ATP), Automatic Train Operation (ATO) and Automatic Train Supervision (ATS) subsystems; and In-Cab Train Advisory (In-CTA) systems which include the In-Cab Information Support System and the Driver Advisory System (DAS). Based on that and the fact that signaling is only a part of TCIS, that is, it represents a kind of TMS in the evaluation just of its impact on railway capacity utilization is not enough. Therefore, in this chapter the evaluation of influence of comprehensive TCIS, as a system composed of various subsystems, on railway capacity utilization is considered.

1.2. Aim and scope of the chapter

Since railway capacity utilization is a multidisciplinary area that depends on different factors, in this chapter applications of MCDM methods for the evaluation of the influence of TCIS on capacity utilization are considered. Based on the fact that signaling as a part of TCIS contributes to efficient railway utilization, the main aim of the chapter is to propose a new approach for evaluation, monitoring and comparison of efficiency of TCIS on both macro and micro levels for improving railway capacity utilization.

The key objective of the chapter is the introduction of the non-radial DEA model for the evaluation of the efficiency of TCIS influence on the improvement of railway capacity utilization. Through the literature review related to evaluation and analysis of railway capacity utilization, application of non-radial DEA model in evaluation of efficiency of TCIS influence on capacity utilization with some real and some assumed data (these were necessary because of the unavailability of the real data and information), only for Serbia and Austria case studies was conducted.

A literature review related to railway capacity analyzing provides confirmation of the novelty of the application of the non-radial DEA model for the efficiency evaluation of TCIS in the improvement of railway capacity utilization and guidelines for selection of desirable and undesirable inputs and outputs for the application of the non-radial DEA model. Also, through application of the non-radial DEA model, sensitivity analysis was presented.

Within the next section, the literature review related to analysis, measurement, evaluation and improvement of railway capacity utilization, followed by an overview of factors and parameters that influence capacity consumption, as well as previous considerations of impacts of TCIS or only signaling systems on capacity utilization are presented. Section 3 contains the history of TCIS and case studies used for the evaluation. Introduction of the DEA method in efficiency evaluation of TCIS on railway capacity utilization with selection of DMUs, inputs and outputs as well as the results of DEA method application are presented in Section 4. Section 5 discusses the application of the MCDM in other areas of railway engineering. Finally, in Section 6 conclusions and proposals for future work are summarized.

2. Literature review

Railway capacity represents an interesting topic and is extensively discussed in the literature by scholars [8]. For railway capacity improvement, two basic approaches are followed: upgrading or expanding the infrastructure, and improving operational characteristics and parameters of the extant rail services. With each approach, it is necessary to assess and analyze the benefits, limitations and challenges through capacity analysis.

There are different capacity analysis approaches and methodologies where input typically includes infrastructure and rolling stock data, operating rules and signaling features [9]. In the following subsections, major comprehensive studies that deal with measuring, analyzing and improving, as well as defining and classifying the capacity utilization, are presented.

2.1. Methods, techniques and approaches in the evaluation of capacity utilization

In the literature, there exist different methods for estimation of utilization of railway capacity and different categorizations of these methods [6]. For example, capacity methodologies have been divided into two major classes as analytical and simulation [10]. Also, capacity methods were categorized to timetable-based and non-timetable-based approaches [11].

Overall, the analytical and simulation methods are the most common methods found in the literature, while one example of combined approach was presented in [12].

A review of the main concepts, methods and techniques of assessing railway capacity was presented in [4, 8, 13], as well as the main factors that can affect railway capacity were also analyzed in [4]. A detailed review of all of these methods, approaches to defining and analyzing and factors affecting the railway capacity may be found in [6].

With these methods in the literature, theoretical, practical capacity as well as used capacity and available capacity were considered [3, 13].

2.1.1. *Analytical methods*

Analytic capacity determination models were developed by [14]. In addition, in [15] also a number of approaches for determining the capacity of railway lines and associated factors in railway operations were discussed. Some of the analytical approaches used in the evaluation of railway capacity for determining absolute capacity could be seen in [16] and others that estimate secondary delays due to route conflicts and train connections in [17]. In [18], an analytical queuing-based approach was proposed while authors of [19] proposed an analytic model. Consideration of capacity evaluation based on the UIC capacity leaflet which is most often used by European railways was presented in [20].

In the UIC code 406 [21], a methodology for capacity evaluation is developed while the applicability of UIC 406 code as a timetable compression technique for evaluation of the corridor and station capacity was considered by [22]. Based on the approach used in the highway capacity manual to analyze capacity of road traffic in [23], a model for railway capacity was developed. Based on that in [24], the headway-based models were further developed.

In [5], an analytical method that is used in Lithuania for calculation of railroad line capacity, *metro-train* method and *UIC 406* method was presented. According to [9], the UIC method is also one of the major tools for improving capacity utilization, whereas in Britain, authors of [25] pointed out Capacity Utilization Index (CUI) also is effective.

Authors of [25] further described the extension of existing capacity utilization measures to enable their application to both junction and station nodes.

A new model for capacity consumption calculation, similar to the UIC 406 method, was presented by [26]. Using the UIC 406 method, [27] calculated the capacity utilization of the Zagreb-Rijeka railway line and sections. Expansion and improvement of the UIC 406 method was made by [28].

2.1.2. *Optimization methods*

Application of an optimization method in analyzing the capacity of new corridors and considering uncertainty of demand levels on the planned route was conducted by [29]. A new method to optimize the travel speed of trains was presented in [30]. Then, in [4] a tool for railway managers called Modulo Optimizador de Mallas (MOM) was developed.

Using an optimization approach in [31], the capacity consumption for railway routes under construction was calculated. [32] presented an integer-programming model for both line and line section for evaluation of train type interactions on railway line capacity. In [33], an optimization approach for analyzing the Beijing-Shanghai high-speed railway line was developed.

An overview of the optimization models used to decide when a fix budget is given and where track duplications and subdivisions should be made has been presented by [34]. In order to solve the station carrying capacity problem with an integer linear programming model was created in [35], while mechanism for optimization of capacity of the main corridor within a railway was developed in [36], while in [37] an optimization algorithm RECIFE-SAT suitable for quantification of capacity by solving the saturation problem adding as many trains as possible to an existing (possibly empty) timetable was proposed.

2.1.3. *Simulation methods*

Simulation methods use general simulation tools, such as AweSim, Minitab and Arena or commercial simulation software specifically designed for rail transportation, such as RTC, MultiRail, RAILSIM, OpenTrack, RailSys and CMS [9]. In [38], Strategic Capacity Analysis for Network (SCAN), developed by [39], was discussed.

A microscopic simulation method for analyzing capacity of a station and complex railway nodes has been introduced by [40, 41], respectively.

In order to analyze railway capacity, Lindfeldt [3] employed several different methodologies.

2.1.4. *Parametric models*

In order to describe and analyze railway capacity, parametric models use some parameters of railway infrastructure. [42] has introduced an enhanced technique of capacity evaluation tools. Parametric models were also developed by [43] for evaluating the capacity of single- and double-track operations using the microscopic simulation software Rail Traffic Controller (RTC). Using RTC, in [44], the simultaneous operations of passenger and freight trains in single- and double-track configurations were compared. Parametric capacity analysis and simulation as two common capacity evaluation methods were described by [45].

2.2. Consideration of influence of TCIS on capacity consumption

Regarding the TCIS and capacity utilization, in [46], a thesis in terms of consideration of capacity increment crossing with the fixed block to the moving-block operation (ERTMS level 3) was presented. In [47], several different methods of capacity evaluation to investigate the effect of different signaling systems on the capacity of a double-track bottleneck in Stockholm were used.

Because advanced signaling systems have recently been considered as a potential solution for increasing system capacity, regardless of the fact that they usually require a great amount of capital investment, the Taiwan Railways Administration (TRA) was interested in estimating the possible capacity benefits of adopting these new signaling systems.

An assessment of maximum capacity at 360 kph (225 mph) based on current technical capability and potential improvements in the capability, were presented by [48].

In order to understand the potential benefits of adopting various advanced signaling systems, such as hybrid or moving-block systems, in [49] a set of analytical capacity models for conventional rail operations with advanced signaling systems, which were implemented in the busiest corridor in the Taiwan Railways Administration system, were developed.

In [50] a new concept of dynamic infrastructure occupation to assess infrastructure capacity was proposed. A new concept was applied in a capacity assessment study of a Dutch railway corridor with different signaling configurations under both scheduled and disturbed traffic conditions. Within their study, comparison of the two different signaling systems, Dutch NS'54/ATB and ETCS Level 2, and their influence on capacity was conducted.

Based on the proposed assessment of the daily capacity improvements deriving from the application of the ETCS signaling system, [51] compared capacity in terms of maximum number of daily trains, between the base Telephone Block system, the ETCS level 2 scenarios and ETCS level 3 with moving block. Using blocking time model, [52] presented headway of signaling systems and headway between trains that are running under different signaling systems. Also, using the simulation software RailSys, they analyzed the effect on capacity when trains are in different signaling systems such as moving block, fixed block and their mix sharing a line of Beijing metro.

In terms of signaling system and capacity, in [53] a design approach able to identify the signaling layout which minimizes the investment and maintenance costs, while respecting the required level of capacity was presented. With increasing the levels of automation within a railway in [54] provided the preliminary hypothesis and methodology for testing them in terms of the capacity increase that comes with them. They concluded that the reliability of a railway network would improve with increasing levels of automation.

2.3. Review of important influence factors and parameters of railway capacity

Since the purpose of this chapter is the evaluation of the influence of TCIS on railway capacity utilization on the macro level, one of the steps of the chapter is presentation of interactions of TCIS with capacity utilization through an adequate set of variables rather than a detailed technical description of different TCIS. Therefore, as a starting point of introducing MCDM methods for evaluation of TCIS in terms of capacity utilization the identification of factors and parameters which affect railway capacity utilization is necessary.

The railway capacity is not static but depends on the infrastructure, operating conditions [4] and overall elements that make up the railway system [55]. As already pointed out, railway capacity is a multidisciplinary area, so different factors affect its utilization [6]. The fundamental factors that affect railway capacity and should be considered in its evaluation, according to [4, 55] are grouped as infrastructure, traffic and operational parameters [4, 55].

Among the infrastructure parameters we can include:

1. *The signaling system* combined with the track layout can be crucial for capacity [3]. A *block section* is a section of track that can only be occupied by one train at a time. *Block section length* is an important factor because together with the spatial distance between two following trains it determines the maximum traffic intensity. For a conventional signaling system with fixed block sections, the length of the block sections on the line is important

for the minimum headway between two consecutive trains in the same direction [3]. Besides the block and signaling systems also, train speed and train length are important infrastructure factors.

2. The important factor for capacity is the *number of tracks on the line*, that is, *single/double or quadruple tracks* [3]. Lindfeldt [3] also pointed out that the capacity of a double track is four times that of a single track, and a quadruple track three times that of a double track. This is possible due to meeting trains everywhere on the line without being restricted to the crossing stations; while on quadruple tracks, trains going in the same direction can preferably be separated according to mean speed. Many until now proposed methodologies only addressed the case of single-track lines with its extension on double-track lines. However, on double-track lines consideration should be separately made for the trains of both directions [55].

3. *The definition of lines and routes* that includes a list of stations and halts along the line and their characteristics is also important because the capacities of railway lines and nodes are key issues in consideration of overall railway capacity [55].

4. *Network effects* that consider what happens on the interfering lines; *track structure and speed limits* that include conditions of the rails, ties and ballast dictate the weight and type of equipment that can be used on the line, as well as the speeds allowed on the line.

   Speed is especially important for single-track lines, where higher average speed means that crossings can occur closer in time, while on double-track lines speed does not have as big an impact on capacity as on single-track lines and has a similar effect for the frequency of overtaking as for crossings on the single-track line. For higher speed in the case that the signaling is based on fixed signal block sections block sections are cleared faster, while the effect is neutralized by trains with higher speed due to longer breaking distance [3].

5. *Length of the subdivision* is lengths of line sections between stations plays important role for railway capacity.

Regarding the traffic parameters the most important factors that should be considered are *new or existing lines; traffic mix* in terms of types of trains and their speeds; *regular timetables; traffic peaking factor;* and *priority* of trains.

Track interruptions; train stop time; maximum trip time threshold; time window and *quality of service, reliability or robustness* represent the most important operational parameters.

One of the important factors for understanding how modern signaling affects and can improve capacity utilization is *blocking time* defined as "the time interval in which a section of track is exclusively allocated to a train and therefore blocked for other trains" [6]. Authors of [27] pointed out that capacity of railway lines depends on the *minimum headway time* of the trains, which is related to the structure of the signaling system; and differences in headway time are the result of changes in travel speed.

According to [8], in the paper [16] it was shown that the capacity of the railway network is affected by factors such as *train speed, train heterogeneity, distance, sectional running times,* and *directional and proportional distribution of trains.*

In [42], they classified *length of subdivision, siding length and spacing, intermediate signal spacing, percentage of number of tracks,* that is, *single, double and multi-tracks, heterogeneity in trains* in terms of train length and power-to-weight ratios among infrastructure and operational factors that influence the railway capacity levels. As key factors that have a direct impact to railway capacity, in [13, 56] *geometrical configuration of the track, line* and *stations* and *layout, features of signaling systems, movement rules* and corresponding *minimum distance between trains,* operation and maintenance planning were pointed out. The main parameters that affect determination of capacity within the infrastructure parameters are *number of tracks, distance between two crossings* or *passing stations, total reference time, theoretical speed* and *project speed,* while operational parameters include *rolling stock features, traffic typology* and *headway* [13].

Among operational effects, parameters that have been included by [13] are generation and propagation of *delays,* as well as required and expected *quality of service.* In order to describe factors that affect railway capacity, used capacity and symptoms of high capacity utilization, in [3] performance indicators based on the available data were selected. Presented indicators for capacity are related to infrastructure, traffic, timetable and delay indicators.

For infrastructure indicators such as *distance between crossing stations (km)* for *single track and distance between passing stations (km)* and *all lines and stations (m)* for *double track* were selected. Within the timetable indicators *number of trains per day (total/passenger/freight), number of trains per hour (total/passenger/freight), speed (km/h) (passenger/freight)* and *speed difference* were selected. In terms of traffic indicators *weight (metric tons), length (m), number of axles* and *axle load* were chosen for freight trains and *gross tons/day (metric tons)* for passenger trains. For passenger/freight, *proportion of trains with increased delay, median of increased delay normalized by route distance [min/100 km]* and *standard deviation of increased delay normalized by route distance [min/100 km]* were placed as delay indicators.

In the model developed in [43], factors used for single-track operation are *siding spacing, signal spacing, track speed, volume (trains/day)* and *heterogeneity.* For double-track, the factors are *crossover spacing,* signal spacing track speed, *volume* and *heterogeneity.* In evaluation capacity of single and double tracks in [44], key factors such as *traffic volume,* defined as *the total number of trains per day (TPD), traffic mixture (heterogeneity)* as the percentage of these that are freight trains and describes the train type heterogeneity of the corridor, *the maximum speed* of the passenger train and the *number of main tracks on the line* were used.

In the literature, the main metrics of capacity level measurements have been categorized through three groups [9]. Within the first group, - i.e., throughput - were categorized *number of trains, tons, train-miles. Terminal/station dwell, punctuality/reliability factor,* and *delay* represent metrics of the second group level of service, and asset utilization, the third group, includes *velocity,* and *infrastructure occupation time* or *percentage.* [9] also emphasized that the Federal Railroad Administration (FRA) introduced a parametric approach developed to measure capacity in the U.S. rail network based on *delay units* (*hours per 100 train-miles*), while European operators use throughput metrics (*number of trains per day or hours*), where punctuality and asset utilization metrics are also applied as secondary units.

In [9] also characteristics that have impact on capacity and its utilization and categorized them into infrastructure, signaling, operational and rolling stock characteristics were reviewed.

However, there is a dilemma as to how these characteristics and their differences were considered in various capacity analysis tools and methodologies.

As demonstrated in literature, the railway capacity rate is controlled by some major components, however in this chapter not all these aspects are taken in account. According to the objective of the chapter, the attention is focused on the evaluation and comparison of the impact of TCIS on capacity utilization deriving from a different TCIS and parameters related to it rather than the other components previously listed.

3. History of the development of railway signaling systems

From 1850, the moving of trains from one to another block was regulated by train dispatchers standing at blocks along the line with a stopwatch. They used hand signals to inform train drivers that a train was going to pass. By 1900, mechanical semaphores were introduced and with the innovation of the telegraph and telephone, the staff was able to send a message (first a certain number of rings on a bell, then a telephone call) to confirm that a train had passed and that a specific block section was finally clear [57].

In 1930, the first optical signals were introduced and the whole system was called *phone block system* [57]. During the middle of the twentieth century, many railways adopted fixed signals. Early fixed signals were semaphores which later on were replaced by traffic light signals with two aspects (green and red), three aspects (green, yellow and red or double green, green and red) or four aspects (green, double yellow, yellow and red) that provided necessary information to the train drivers [6].

3.1. Introduction of the semiautomatic and automatic block systems

The *semiautomatic block* was introduced when fixed signals had replaced hand signals. However, today railways are equipped with signaling systems based on the automatic blocks which do not require manual intervention. With automatic blocks (interlocking), the rail line is divided into blocks of which the length should be longer than the braking distance of the faster train running on the route, and where the occupancy of the blocks is detecting with the train detection equipment (track circuit and axle counter). In the early 1980s, railway signaling systems were upgraded in order to be able to constantly monitor the speed of the train and improve safety, through automatic braking if the driver fails to respond to the warnings such as a train passing a red (danger) signal or exceeding a speed restriction – these are known as Automatic Train Protection (ATP) systems [57]. Each country has its ATP, and there are now 20 different developed and operated ATP systems in the EU [58] according to the different national requirements, technical standards and operating rules [57].

3.2. Development of standard European signaling system

In order to remove more than 20 different TCISs across the European railway network [58], the EC decided to support the implementation of a harmonized standard for the railway control and signaling systems, the European Railway Traffic Management System (ERTMS) [59]. The ERTMS as a kind of Traffic Management System (TMS) is one of the classes of TCISs along with the Interlocking Systems (IXL), Automatic Train Control (ATC) systems and Automatic Train Supervision (ATS) systems [7]. The study [7] gives detailed classification of TCISs classes. ERTMS as a standard Automatic Train Protection (ATP) and Automatic Train Control (ATC) are complementary systems that ensure safer operation of trains and can enhance capacity utilization [6].

Besides standardization, ERTMS was created to providing interoperability for different countries and manufacturers leading an increase in the efficiency of traffic of both passenger and freight trains and the competitiveness of the European railway [60]. The realization of interoperability is a precondition for other benefits of ERTMS, such as improved traffic management, optimized usage of energy and network resource [61], enhanced profitability and customer service, with a contribution to overall environmental and energy efficiency objectives [62]. In addition, higher levels of safety for high-speed trains with only a few minutes headway between trains ensures operational reliability and more traffic within the same infrastructure, that is, better capacity utilization, reduction of maintenance costs [60, 63, 64] as well as the creation of a common market of signaling equipment that also leads to reduction of costs [62–64].

3.2.1. Description levels of ERTMS

ERTMS consists of three levels. Level 1 represents track to train communication, provided by Eurobalises located along tracks that interface with the existing signaling system and line-side signals. It is based on 'fixed block' and fixed braking distance. However, the lower deceleration rate in ERTMS level 1 results in lower capacity utilization.

Level 2 is an enhancement of Level 1 with movement authority from Radio Block Centre (RBC) through interlocking to on-board ECTS via GSM-R link, which enables the elimination of trackside signals. Level 2 covers track-to-train communication and vice versa. On this level, RBC calculates the correct movement authority, giving authorization to proceed (or not), with balises used to transmit static messages such as location, line profile and speed limit. This level enables higher operational speeds and reduced headways, thus more efficient capacity utilization because it is also based on a 'fixed block system' though it eliminates the time needed for the driver to see the trackside signals as there is cab signaling which is displayed on-board.

Level 3 improves the ability of Level 2 so that train detection by the trackside is no longer required. At this level, the RBC uses GSM-R for transmission between track and train. Compared with Level 1 and 2 that are based on a fixed block signaling system, Level 3 allows a moving-block signaling system; this means that as the train travels, the track receives the train location and train integrity from the train. Levels 1 and 2 are already widely applied in Europe, while Level 3 is currently under development [6]. Therefore, in modern signaling, like an ERTMS Level 3, line-side signals are not required as the necessary information is displayed in the driver's cabin and intends to introduce moving blocks which can further decrease headways [6]. With ERTMS level 3, the time needed for the driver to see the trackside signal because of cab signaling (because moving blocks are dynamic and can be much shorter than fixed blocks) is eliminated. Hence, with European Cab-based signaling (ERTMS) improvement of the capacity utilization can be significantly achieved by more efficient train control and higher speed while safety can be also increased through reducing the safety distance, that is, shorter blocks between trains and decreasing headways so the time of the occupied block results in faster journeys [6].

3.3. Case studies of TCIS evaluation

Within the EU each country has its own TCIS. For introduction of MCDM in evaluation of the influence of TCIS on capacity utilization based on the available data and information used for assumption of data only two case studies, that is, Serbia and Austria were considered.

3.3.1. Austria case study

Austrian TCIS: The almost comprehensive Austrian railway network has a computerized train monitoring system. The train radio system was upgraded to European GSM-R standard by infrastructure managers of the Austrian railway network. Further, the Austrian railway network is equipped with the "Punktformige Zugbeeinflussung" system (intermittent automatic train running control system, PZB), which is the descendant of the "Linienzugbeeinflussung" system (continuous train protection, LZB) and provided warnings and enforced train braking.

Also, on some rail sections LZB and ETCS systems are in use. Newly constructed lines "Wienerwaldtunnel" (Abzweigung Knoten Hadersdorf – Tullnerfeld – Knoten Wagram) and "Unterinntal" (Kundl – Radfeld – Abzw. Baumkirchen), planned to be equipped with a reduced level of signaling, can be used exclusively by trains equipped with the Level 2 ETCS, while all other lines equipped with Level 1 ETCS can also be used by the PZB system.

On the Austrian network, ETCS Level 1 system was installed on five rail sections of total length 381 km, while ETCS Level 2 system also in five rail sectors of total length of 447.4 km. Several plans exist for investments concerning the installation of ETCS Level 1 and 2 systems which are expected to be ready through a time period from 2012 to 2025 [65].

3.3.2. Serbian case study

Serbian TCIS: On the Serbian railway network TCIS differ from one railway line to the other. Along the railway line Belgrade – Novi Sad – Subotica a copper signal/telecommunication cable is used and on this railway section an automatic railway network is in function. The telecommunication equipment installed is at very low level, while the electrical system was brought to its limit for the railway line Belgrade-border Croatia.

Along the railway line Belgrade – Niš and in specific sectors, an STA signal – telecommunication coaxial copper cable is in use. On the railway line Niš – Preševo, a signal – telecommunication copper STA cable is in use and the line is covered by a radio dispatch system for direct communication between the telecommand center in Niš and trains in operation.

The telephone switching system operates by means of switching nodes of outdated technology based on space division multiplexing. On these two lines, the railway telephony system and dispatch links were realized in relay technology from the 1940s. Along the railway line Niš-border of Bulgaria, there are trackside telecommunication devices and aerial telephone cables installed and used. In Dimitrovgrad, there is an automatic railway operation network terminal (so-called ŽAT) center which is connected to an existing nodal ŽAT center in Niš. None of these systems support the installation of modern digital signaling devices, systems for traffic control and information technologies. Based on this, it is clear that the Serbian railway network is equipped with outdated telecommunication systems.

On the Serbian network, there is no automatic traffic control, but an intermittent inductive automatic train stopping device for the control of the trains, including the stationary part close to the track, transmission system and locomotive part. Regarding the ETCS/ERTMS at Serbian railway along the lines of Corridor X are planned to be installed [65].

4. Methodology of the TCIS evaluation with non-radial DEA model

Within the first step of this chapter, the non-radial DEA model was introduced. After that for the purpose of evaluation of TCIS inputs and outputs for the introduced model (M) were defined and then these inputs and outputs were classified as desirable and undesirable. Since the overall real data is missing, application of the non-radial DEA model was tested mainly on the assumed data. This step was followed by a selection of DMUs for evaluation of TCIS, that is, years of TCIS functioning for case studies, and based on the availability, data for them were collected. In the fourth step, the selection of weights for desirable and undesirable inputs and outputs was performed. Finally, the validation of the model (M) through perturbation of the used data was presented in order to check the behavior of the model. In the following subsections, the detailed description of all steps are given.

4.1. Introduction of DEA method in the evaluation of efficiency of TCIS on railway capacity utilization

As the objective of this chapter is to determine the efficiency of TCIS and the monitoring of its effects on railway capacity utilization, there is a need to provide a potential tool for monitoring, evaluating and benchmarking changes of TCIS impacts on the consumption of railway capacity. In this study, the Data Envelopment Analysis (DEA) methodology, as a good tool for measuring the efficiency and impacts of TCIS in terms of efficiency of using railway capacity at different levels, was considered.

DEA can be used for tactical and strategic planning especially where managerial comparisons between the relative efficiency of some units (e.g., railways of different countries, train-operating companies, stations, etc.) [66]. The DEA is a well-known "non-parametric productive efficiency measurement method for operations with multiple inputs and multiple outputs" [67].

The DEA method, first popularized by [68], combines and transfers multiple inputs and outputs into a single efficiency index, forming the "efficient frontier" with a set of Decision-Making Units (DMUs) pointing toward best practices and assigning a level of efficiency to other DMUs that are not on the frontier according to the distance to the efficient frontier [67].

Today, there are different DEA models for measuring efficiency for different types of measuring requirements [67]; among them, according to [69], the most well-known are the CCR model [68], the BCC model [70], the Additive model [71] and the Cone Ratio model [72]. Within these DEA models, only desirable inputs and outputs are considered, without consideration of undesirable inputs and outputs which can appear in real applications. For example, in evaluating the efficiency of TCIS on railway capacity, different factors can be modeled as undesirable. Therefore, in this study, non-radial DEA model which modeled some inputs and outputs as undesirable were introduced and considered.

DEA has been used by economist in the field of railways in order to analyze economic efficiency. However, it was also applied to railway capacity and provided promising results on improvement of the capacity utilization at railway stations and passenger train operators in the United Kingdom [66].

For the measuring of efficiency related to the contribution of TCIS on better utilization of railway capacity on the macro level, non-radial DEA model (M) that includes both desirable and undesirable inputs and outputs was employed. Based on the data for Serbia and Austria, the non-radial DEA model (M) was piloted for evaluation of efficiency of TCIS contribution to railway capacity. In the investigation of the non-radial DEA model (M), factors that affect the railway capacity utilization and are related to the TCIS were considered as guidelines in the selection of inputs and outputs. Besides the other widely used non-radial DEA models such as Slack-based models, Russell measure models and Directional distance function, in this chapter non-radial DEA model (M) was chosen due to its ability to employ different non-proportional adjustments, decision-maker specified weights assigned to each efficiency score, and because of its ability to proportionally decrease the amounts of energy inputs and undesirable outputs simultaneously as much as possible [73, 74].

Based on the results obtained by applied data and sensitivity analysis, it could be said that the proposed non-radial DEA model (M) could be applicable for providing information about the role and efficiency of TCIS on using railway capacity for various levels. Moreover, it could be used as a support tool for identifying the best practices in terms of the type of TCIS, factors that cause better or worse roles of TCIS related to use of railway capacity, as well as testing and monitoring the role of TCIS in different railway capacities such as theoretical and practical.

4.2. In brief about DEA model

DEA is a linear programming method for efficiency measurement based on Farrell's original work [75] that was later popularized by authors of [68]. DEA has been commonly applied in empirical literature as a non-parametric frontier approach for evaluating the relative efficiency of a comparable set of entities, called DMUs, with multiple inputs and outputs, that is, DMUs that are able to transform multiple inputs into multiple outputs. This method offers decision-makers (DMs) information on efficient (i.e., best practice) and non-efficient DMUs. A major stated advantage of the DEA is that it does not require any prior assumptions on underlying functional relationships between inputs and outputs [76]. Additionally, the method does not require prior definition of weights of criteria for input and output by DMs, as all weights are determined after solving the DEA model, which eliminates subjective decision-making. Assuming that there are n DMUs, m outputs and s inputs, efficiency score is usually calculated based on the input-oriented Charnes, Cooper and Rhodes (CCR) DEA model [76, 77] that can be written as: $\min\theta; s.t\, X\lambda \leq \theta x_i, Y\lambda \geq y_i, \lambda \geq 0$,

where X, Y and x_i, y_i are data matrices and vectors of inputs and outputs, respectively, λ is a vector of variables. θ represents an indicator of technical efficiency, where $\theta \in [0, 1]$ and indicates how much an evaluated entity could potentially reduce its input vector while holding the output constant. The presented CCR model exhibits the constant returns to scale (CSR), but with the additional constraint $\sum\lambda = 1$, CCR model becomes the classical Banker, Chames and Cooper (BCC) model that allows the variant to return to scale (VRS) [70, 76].

4.3. Description of non-radial DEA model

The classical DEA model is strongly related to, and can be presented through, production theory, where raw materials and resources are treated as inputs, while products are treated as outputs in the production process. However, in some real applications, production process may also use undesirable inputs and generate undesirable outputs [69], like smoke pollution or waste [78, 79]. Consequently, authors of [80, 81] have stated that in that production process both desirable and undesirable factors may be presented within the models of DEA methodology.

DEA models with undesirable inputs and outputs have been extensively studied [69]. Some of these papers are summarized below. For instance, in [81] an alternative approach to treating both desirable and undesirable factors differently in the BCC model was developed. In [82], the control of changes in input/output levels of a given DMU with the presence of undesirable factors in order to preserve the efficiency index of a DMU was considered. In [79], a model in the framework of DEA for treating undesirable inputs and outputs was proposed. A method for treating both undesirable inputs and outputs simultaneously in non-radial DEA models was presented in [83]. Furthermore, in [80] a CCR-DEA model was extended to a DEA-like model able to deal with undesirable inputs and outputs. In [69], authors have discussed a general approach of deriving DEA models to handle undesirable inputs and outputs without transferring undesirable data, such as in [81]. Including undesirable inputs and outputs, DEA has been extensively represented and used in environmental fields, such as energy and environmental efficiency. Additionally, a large number of papers have focused on the evaluation of transport energy or environmental efficiency [73].

Considering the efficiency of TCIS on utilization of railway capacity through the transportation process, undesirable outputs of TCIS functioning could appear, such as failures of the system. If inefficiency is present, the undesirable output should be reduced to improve inefficiency, that is, the role of TCIS on utilization of railway capacity, which means that undesirable and desirable outputs should be treated differently when we evaluate the impacts of TCIS on efficient railway capacity use. Based on the paper [84] regarding energy efficiency, production process, desirable and undesirable outputs are jointly produced by consuming both desirable and undesirable inputs, where x, e, y and u are vectors of energy inputs and non-energy inputs (i.e., here undesirable inputs and desirable inputs), desirable outputs and undesirable outputs, respectively. Therefore, the joint production process can be represented as $T = \{(x, e, y, u) : (x, e) \text{ can produce } (y, e)\}$. Based on this T reference technology, radial model, modified-radial and non-radial models such as the Russell measure model, Tone's slack-based model, range adjusted model and directional distance function model are used in energy efficiency and carbon emission efficiency in literature. Additionally, there are four types of returns to scale (*RTS*) such as constant *RTS* (*CRS*) which is the most used *RTS* category, non-increasing *RTS* (*NIRS*), non-decreasing *RTS* (*NDRS*) and variant *RTS* (*VRS*), where each of them reflects reference technology [73].

The radial model presented in [84] aims at reducing energy inputs as much as possible for the given level of non-energy inputs, desirable and undesirable outputs. Since the radial model has weak discriminating power in energy efficiency comparisons and does not consider energy mix effects, non-radial models for energy efficiency evaluation is also proposed in [84, 85]. They have presented application of non-radial DEA models for energy efficiency evaluation considering undesirable outputs and maximized energy-saving potential, all under CRS, NIRS and VRS. For example, if in model (M) instead of limitation (5) we write $\sum_{k=1}^{K} \lambda_k \leq 1$, $\sum_{k=1}^{K} \lambda_k \geq 1$ ili $\sum_{k=1}^{K} \lambda_k = 1$ we receive non-radial model under NIRS, NDRS and VRS, respectively. However, their non-radial models also attempt to reduce energy inputs as much as possible for the given level of non-energy input, desirable and undesirable outputs. In other words, their non-radial models do not consider reduction of undesirable outputs.

Furthermore, radial and non-radial DEA models for evaluating DMUs total-factor *energy and environmental efficiency* have been presented in [74]. In order to overcome the discriminating power of the radial model, following [86, 87] in [74], the radial DEA model for energy and environmental efficiency evaluation has been extended to a non-radial model. Assuming that there are K DMUs, and each DMU has n desirable inputs (non-energy inputs) and l undesirable inputs (energy inputs) in order to produce m desirable outputs and j undesirable outputs denoted, respectively, as $x = (x_{1K}, \ldots, x_{nK})$, $e = (e_{1l}, \ldots, x_{LK})$, $y = (y_{mK}, \ldots, y_{mK})$, $u = (u_{1K}, \ldots, u_{JK})$, the non-radial DEA model denoted (M) is the following:

$$min \frac{1}{2}\left(\frac{1}{L}\sum_{l=1}^{L}\theta_l + \frac{1}{J}\sum_{j=1}^{J}\theta_j\right)$$

s.t.

$$\sum_{k=1}^{K} \lambda_k x_{nk} \leq x_{n0}, n = 1, \ldots..N \tag{1}$$

$$\sum_{k=1}^{K} \lambda_k e_{lk} \leq \theta_l e_{l0}, l = 1, \ldots\ldots.L \tag{2}$$

where

$$\sum_{k=1}^{K} \lambda_k y_{mk} \geq y_{m0}, m = 1, \ldots.M \tag{3}$$

$$\sum_{k=1}^{K} \lambda_k u_{jk} = \theta_j u_{j0}, j = 1, \ldots .J \tag{4}$$

$$\lambda_k \geq 0, k = 1, \ldots .K. \tag{5}$$

This non-radial model (M) could be projected for evaluation of efficiency of TCIS in utilization of railway capacity. When compared to other non-radial models mentioned in the literature, the non-radial model (M) proportionally decreases the number of undesirable inputs and undesirable outputs as much as possible for the given level of desirable inputs and desirable outputs. The optimal values **of unified efficiency** are in the interval between 0 and 1. An entity with a higher value of efficiency has better efficiency in terms of the degree of the role of TCIS on capacity consumption as compared to others entities.

In the case of the non-radial DEA model (M), if the entity has an objective function equal to 1 it means that the entity is the best, located on the frontier, and could not reduce undesirable input and undesirable output. Such a non-radial model (M) could be applicable for evaluation of the efficiency of TCIS in railway capacity utilization because it has a relatively strong discriminating power and capability of expanding desirable outputs, while simultaneously reducing undesirable outputs. Additionally, **unified efficiency** can be calculated through decision-maker specified weights assigned to each of these two efficiency scores and depends on the preferences between undesirable inputs utilization and undesirable outputs. In the model proposed in [74], the weights were set to 1/2. Hence, besides additional widely used non-radial DEA models such as Slack-based models, Russell measure models and Directional distance function, in this chapter non-radial DEA model was chosen due to its ability to use different non-proportional adjustments, decision-maker specified weights assigned to each efficiency score, and because of its ability to proportionally decrease the amounts of inputs and undesirable outputs simultaneously as much as possible [73, 74].

4.4. Selection of case studies/DMUs and variables

In order to apply the non-radial DEA model, one of the most important steps is the selection of inputs and outputs, as well as classification of them as desirable and undesirable. Also, in [51], it was pointed out that the first point in assessing the performance improvements deriving from a better signaling system is to set what values have to be chosen for comparison.

One of the most traditional measures in railway capacity evaluation is the number of trains that travel in each corridor in a given time. However, in practice, there are different metrics that are equally important and could be used in capacity analysis [8]. Also, capacity should be analyzed in relation to the railway's primary mission, which is the transportation of freight and passengers [88]. Consequently, capacity can be evaluated through measures of network performance such as number of trains, number of wagons, number of pieces of cargo, number of passengers.

As can be seen from the literature review, different merits can be applied for evaluation of impact of TCIS on the railway capacity. However, mainly based on the available data, only limited merits were selected within this chapter. Therefore, based on the above reviewed factors that affect railway capacity, data availability and the fact that railway capacity can be described by other indicators, for the purpose of evaluation efficiency and influence of TCIS on capacity utilization, desirable and undesirable inputs and outputs were proposed. First, because railway transportation can be represented as a production process, as desirable inputs, *length of railway network (DI1)* and *number of trains (per day) (DI2)* on the railway network were selected because they represent timetable indicators. As outputs of production process of railway transportation realized *freight kilometers (DO1)* and *passenger kilometers (DO2)* [88] were included in non-radial DEA model as desirable outputs. It is clear that higher value of DO1 and DO2 capacity was increased which is happening on the liberalized railway markets. Therefore, capacity is the maximum amount that can be produced in relation to the limiting constraints from infrastructure, rolling stock or staff [88]. Consequently, as undesirable input, which is close to the functioning of TCIS, *number of failures of the whole system or its subsystem (UDI)*, was proposed. Then, as an undesirable output, *punctuality of the trains (UDO)*, which is the result of system failures, was selected.

A second step of DEA methodology is the selection of comparable DMUs. For the application of the non-radial DEA model and consideration of the results of the model, based on the data availability for defined inputs and outputs, Serbia and Austria with their own TCIS as case studies were selected.

DMUs of selected case studies represent years. For each Serbian DMU, real data were used. Data for indicators such as *number of trains (per day)* and *punctuality of the trains* were collected from planned and realized timetables. For the *number of failures*, data were collected from the evidence of Serbian railways, while realized *freight and passenger kilometers* and *length of railway network* data were extracted for Serbian statistics.

Real data for the Austria case, published by OBB [89] were used only for 2015 and were collected for *length of railway network* and *number of trains (per day)* while data for *freight and passenger kilometers* Eurostat was used. However, data for *number of failures* were assumed because of missing data for that indicator. Data for other years were assumed due to unavailability of data for each year. All data used can be seen in **Table 1**.

	Serbia case study						Austria case study					
DMUs	**DI1**	**DI2**	**DO1**	**DO2**	**UDI**	**UDO (%)**	**DI1**	**DI2**	**DO1**	**DO2**	**UDI**	**UDO (%)**
2006	3819	1510	684,110	4232	55	40	9646*	6327*	110,778	8907	8	90*
2007	3819	1515	687,002	4551	43	55	9646*	6329*	115,526	9167	7	95*
2008	3819	1502	583,071	4339	38	60	9646*	6345*	121,579	10,365	7	95*
2009	3819	1430	522,033	2967	35	65	9646*	6332*	98,887	10,184	9	80*
2010	3819	1431	521,933	3522	39	60	9646*	6340*	107,670	10,263	10	85*
2011	3819	1431	540,911	3611	34	70	9646*	6340*	107,587	10,778	7	95*
2012	3819	1430	539,727	2769	23	80	9646*	6339*	100,452	11,211	6	96*
2013	3819	1433	612,495	3022	34	70	9646*	6330*	95,449	11,804	8	90*
2014	3819	1420	452,963	2988	27	80	9646*	6335*	98,281	11,981	7	95*
2015	3739	1436	508,678	3249	30	80	9646	6340	97,642	12,104	5	96.3

*Denotes assumed data.

Table 1. Data used for non-radial DEA model.

4.5. Results of non-radial DEA model and sensitivity analysis

With the aim to present the applicability of the non-radial DEA model in measuring the efficiency of TCIS on railway capacity utilization, using the real and assumed data, results of the application through case studies are presented in **Table 2**. All the results were obtained using Excel Solver. Based on them, it could be concluded that the non-radial DEA model can be applied for evaluation efficiency and influence of TCIS on railway capacity utilization. Results represent efficient and non-efficient DMUs, that is, years in which the influence of TCIS on capacity utilization was higher or lower. Since the real data were collected only for the Serbian case study, the results are valid. On the contrary, the results of the Austrian case study are not valid because we used the real data only for 1 year, and all other data were assumed. Therefore, the second case study was used only as an example in order to check the applicability of the non-radial DEA model.

However, in order to check the stability and behavior of the non-radial DEA model, the sensitivity analysis was conducted for each case study (for results see **Table 2**). Consequently, the ideal way, which can be related to sensitivity analysis, is therefore to vary some parameters while monitoring the relative movement of efficiency changes [51]. Like, for example, to vary the indicators, that is, UDI and UDO – while observing the results of the influence of TCIS on railway capacity consumption. Therefore, the sensitivity analysis was conducted through data variations applying "approach II" through "development model" [90]. Sensitivity analysis was made for certain percentages of perturbation (10, 20 and 50%) until the status of at least one DMU was changed from inefficient to efficient or vice versa [77]. For efficient countries, both undesirable inputs and outputs were increased and simultaneously decreased for the inefficient one. After each data perturbation, the calculation performed by Excel Solver and the results obtained are represented in **Table 2**. From the results, it can be

	Serbia case study				**Austria case study**			
	Results of non-radial DEA model	**Sensitivity analysis of non-radial DEA model**			**Results of non-radial DEA model**	**Sensitivity analysis of non-radial DEA model**		
DMUs		**10%**	**20%**	**50%**		**10%**	**20%**	**50%**
2006	1	1	1	1	0.8795	0.928	0.928	0.928
2007	1	1	1	1	0.9502	1	1	1
2008	0.9764	1	1	1	1	1	1	1
2009	0.7882	0.8807	0.8807	0.8807	0.8412	0.898	0.898	0.898
2010	0.7813	0.8795	0.8795	0.8795	0.8296	0.8871	0.8871	0.8871
2011	0.8134	0.9059	0.9509	0.9505	0.9107	1	1	1
2012	1	0.8531	0.6951	0.3475	0.9274	1	1	1
2013	0.914	1	1	1	0.8279	0.951	0.951	0.951
2014	0.7516	0.8215	0.8215	0.8215	0.8684	1	1	1
2015	0.7851	0.8604	0.8604	0.8604	1	1	0.9999	0.9999

Table 2. Results and sensitivity analysis of application of non-radial DEA model.

concluded that the efficiency score was improved for all inefficient years. With 10 and 20% of decrement in data for the Serbia, firstly inefficient years 2008 and 2013 became efficient, while the efficiency score was changed for efficient year 2012 after the 10% of increment in data. In terms of the Austrian case with 10% of decrement in data inefficient DMUs, that is, years 2007, 2011, 2012 and 2014 – became efficient, while for 2015 the efficiency score was changed with 10% of increment in data.

Taking into account the results of sensitivity analysis, it can be noted that sensitivity analysis shows some anomalies or weaknesses in the non-radial DEA model. The first is related to the sensitivity of the efficiency score with a smaller percentage of modifications in undesirable input and output. In the case of inaccurate data, the model can provide an unrealistic picture regarding the best DMUs. Another weakness of the model can be related to the correctness of the results depending on the number of variables included. It should be noted that with a higher percentage of data modifications the model would provide a picture related to more significant changes of the efficiency score. Also, it is clear that without an adequate classification of variables as inputs and outputs, results can be inaccurate.

5. An overview of the applications of MCDM in railway engineering

Based on the searches performed in Scopus and Science Direct databases, in the literature application of different MCDM in railway engineering such as AHP, ANP, DEA, TOPSIS, VIKOR and DEMETAL can be found, while DEA, used in terms of capacity evaluation, can be found also in other fields of railway. A detailed literature review related to application of DEA in the railway can be found in [6]. These methods were used alone or in combination with other MCDM. Also, in the literature some multicriteria hybrid approaches in railway engineering could be found. For solving multicriteria decision problems such as railway capacity, in [91] the RECIFE software as a multicriteria decision supports system for the evaluation of railway capacity at the station or junction level.

5.1. Analytic hierarchy process (AHP)

In the literature, the analytic hierarchy process (AHP) appears as one of the most commonly used MCDM. According to [92], the major multi-criteria decision methods used for dealing with railways management problems were AHP and ANP. AHP is used also to get indicators' weight values in [93], for determination of weights of each index [94], in evaluation of the railway infrastructure objects from the perspective of traffic safety risk [95] and to find the weights for the selection criteria in selecting the most appropriate supplier [96]. Through installation of the Train Conformity Check system (TCCS) that can detect and alert dispatchers about several dangerous or damaging defects on rolling stock [97] have applied AHP to evaluate the optimal locations to install a TCCS on a railway section.

In order to identify the most suitable manufacturer of rail vehicles for the UK infrastructure project, High Speed 2 in [98] applied AHP for effective comparison of the four primary rolling stock manufacturers: Bombardier, Siemens, Hitachi and Alstom.

In improving the public transport system of Istanbul through the planning of the rail transit network in [99], AHP was used in evaluation of the final three alternatives. In terms of the quality of railway passenger transportation, AHP was used for determination of weights for criteria [100] and indictors [101, 102] and for the different maintenance actions in the railway infrastructure [103], relative weights of quality of service criteria in the high-speed railway wireless communication networks [104], weights of 16 criteria of the quality of the trip by train [105]. In [6], AHP was used to develop a model for evaluation of TCIS and their subsystems, as well as KPTs in terms of sustainability.

Besides the application of traditional AHP in the railway field the papers that aimed at modeling the fuzziness in the AHP which allows inclusion of human experts and their communication of linguistic variables [106] are found. In combination with the fuzzy set theory, Li [107] has applied AHP to select the best design scheme of the railway freight car. In [108], the Fuzzy AHP was used to develop a risk assessment system for evaluating both qualitative and quantitative risk data and information associated with the safety management of railway systems. A hierarchical customer satisfaction framework was made by [109] where fuzzy AHP has been applied for calculation of weights of main criteria. After application of fault tree analysis to analyze historical general accidents in railway transportation in [110], AHP was employed to analyze relationships between the factors and general accidents. Group fuzzy AHP was used by [111] for ranking and selection of key performance indicators of railway control rooms.

5.2. Analytic network process (ANP)

The adoption of ANP for the evaluation of maintenance strategy for rolling stock of railway system operators formed by various combinations of preventive maintenance (PM) and corrective maintenance (CM) was presented by [112]. Using ANP in [113], the methodology was developed and applied to a technical maintenance project for the Spanish National Railway Infrastructure company for measuring stakeholders' influences within a project. For the historic Alishan Forest Railway in Taiwan, in [114] the analytic network process (ANP) was used in evaluation of different revitalization strategies based on the various factors and the interactions of those factors.

5.3. Technique for order preference by similarity to ideal solution (TOPSIS)

In evaluation of high-speed transport systems where alternatives are represented as High-Speed Rail and Transrapid Maglev, TOPSIS was applied in [115] for selection of the preferable alternative. The TOPSIS method with multilevel grey evaluation (MGE) was employed by [116] to evaluate the overall performance of passenger transfer at large transport terminals. Fuzzy TOPSIS with failure mode and effect analysis was proposed by [117] for determination of the closeness coefficient of each failure mode of metro door fault criticality. For measurement of a service quality of rail transit lines in combination with statistical analysis, trapezoidal fuzzy numbers has been adopted, Fuzzy-TOPSIS [118].

5.4. Data envelopment analysis (DEA)

In order to recognize changes in the efficiency and productivity of railway freight transportation in Europe, [119] applied DEA to analyze different European countries from 1980 to 2003.

For analyzing and comparing the synthetic indices of accessibility for evaluation impacts of high-speed trains on European cities [120] considered DEA and Principal Component Analysis (PCA) which were employed in the their previous two studies. Using a two-stage bootstrapped data envelopment analysis (DEA) with incorporation of transaction costs in order to assess the desirability of vertical separation in [121] authors compared 43 Swedish, German and British train operating firms.

The methodology for defining, measuring and analyzing the capacity utilization by DEA in the passenger railway operation sector and freight sector was developed by [6]. A DEA cross evaluating method based on grey incidence analysis was adopted by [122] to evaluate and estimate the efficiency of shunting locomotives operation in train service depots.

For evaluation of the inter-regional railway operation performance for 30 provinces in China, [123] adopted the SUPER-SBM DEA method dealing with uncontrollable factors. By DEA, in [66] the relative operational efficiency of 24 European railways in capacity utilization was studied.

Throughout the investigation of the potential impacts on rail accessibility across the Europe for different scenarios, in [119] the DEA method was employed. Working on the real-time optimization of train scheduling decisions at a complex railway network during congested traffic situations, [124] used DEA in evaluating the relative efficiency of the different optimization formulations.

In order to use an adequate approach to quantify and rank the relative performance and efficiency of stations, in [125] two novel models of the DEA method as a new tool for analyzing macro and micro capacity utilization at stations were suggested. In order to measure the success of the liberalization process of railway transportation in [126], DEA was used for estimation of technical effectiveness of railway performance; [127] used Fuzzy DEA to evaluate the efficiency of the rescheduled timetable in terms of delay minimization and robustness maximization.

5.5. VIKOR and ELECTRE methods

In the methodology of route selection applied for railway route planning and design in [128], Vlsekriterijumska Optimizacija I Kompromisno Resenje (VIKOR) was used for the selection of the most favorable railway route. Within the framework developed for measuring the level of customer satisfaction related to rail transit network (metros, trams, light rail and funicular) in Istanbul, in combination with other techniques in [129] applied VIKOR with interval type-2 fuzzy sets to obtain the best customer satisfaction level of rail transit network based on average and the worst group scores among the set of alternatives. VIKOR, with an intuitionistic fuzzy set, was also introduced and employed by [130] for CRH2 high-speed train bogie system operation safety assessment.

Based on the literature review related to multicriteria decision analysis in transportation, through using a screening method and ELECTRE I, in [131] selection of a high-speed rail corridor/route that can be useful in planning high-speed rail in Malaysia was provided.

5.6. Hybrid/combined MCDM approaches

In the literature some combinations of MCDM approaches in railway engineering could also be found. For example, in [132] authors integrated DEA and AHP with computer simulation for railway system improvement and optimization. Using AHP and DEA, a selection of the optimum site for a railway station was presented in [133]. In [134], the evaluation of world railways performance with a fuzzy dynamic multi-objective DEA model was presented.

In the two-phase model for ranking railway projects from different railway subsystems in [135], TOPSIS was used with AHP in [136] an improved Dempster-Shafer theory (DS)/AHP method for the evaluation of hazard source was provided. For the purposes of evaluation and selection of optimal locations for freight villages, authors of [137] developed a GIS-ANP-TOPSIS framework. For selection of the most appropriate maintenance policy, an integrated MCDM approach was developed by [138] on the basis of FANP technique and in combination with the DEMATEL. In [139], ANP and DEMETAL were used to determine the leading accident casual factors and to analyze the influence of the relationships of human and organization factors.

6. Conclusions and future work

This chapter focused on the introduction of a new approach in efficiency evaluation of TCIS regarding the influence on improvement of capacity utilization. For the purpose of efficiency evaluation of TCIS, the non-radial DEA model was proposed and employed. The efficiency evaluation of TCIS influence on improvement of the capacity utilization with the non-radial DEA model was conducted by including desirable and undesirable inputs, as well as desirable and undesirable outputs. The evaluation was tested through two case studies, that is, Serbian and Austrian for the period from 2006 to 2015 mainly based on the assumed data for the Austrian case, where years represent DMUs. The results of the non-radial DEA model showed the best efficiency year in terms of the influence of TCIS on capacity utilization, as well as those with low(er) efficiency.

Based on the performed sensitivity analysis, it can be said that the non-radial DEA model is valid and can be applicable for the evaluation efficiency of TCIS influence on capacity utilization for different levels, that is, overall network or particular line. However, there is a significant sensitivity to data for a smaller data variation that cause reduced stability of the model. In the case of inappropriate and missing data, results can be different – it is certain that results of the model will be different because sensitivity analysis showed instability with a lower data variation. Overall, these weaknesses can affect the results of the non-radial DEA model and provide a thankless picture regarding efficiency in terms of the influence of TCIS on railway capacity utilization. However, in the case of accurate data and selection of appropriate inputs and outputs in accordance with the aim of the evaluation, the proposed new methodology could be a good tool for efficiency evaluation of TCIS influence on capacity utilization.

Through empirical study, it can be shown that the non-radial DEA model, employed using an accurate set of data, can provide an evaluation of the efficiency of the TCIS influence on capacity utilization both at the macro and micro levels, as well as benchmarking for different levels. The evaluation of other factors and their impacts of TCIS on capacity utilization are also possible with the non-radial DEA model.

As future work, based on a comprehensive and accurate set of data, the proposed methods can be used with different variables/criteria and tested in order to check their validity. Also, as a part of the future work, methods can be applied on the micro level for evaluation of the influence of TCIS on capacity utilization for a particular line. Moreover, with this MCDM, the evaluation, measuring and comparison of railway capacity consumption for different concepts of capacity can be conducted.

Author details

Boban Djordjević and Krmac Evelin*

*Address all correspondence to: evelin.krmac@fpp.uni-lj.si

Faculty of Maritime Studies and Transport, University of Ljubljana, Slovenia

References

[1] European Commission. Commission Staff Working Document: A European Strategy for Low-Emission Mobility. Brussels: European Commission; 2016

[2] UIC/CER. Rail Transport and Environment: Facts and Figures. Paris; 2015

[3] Lindfeldt A. Railway capacity analysis: Methods for simulation and evaluation of timetables, delays and infrastructure. [Doctoral thesis]. Stockholm: KTH Royal Institute of Technology; 2015

[4] Abril M, Barber F, Ingolotti L, Salido MA, Tormos P, Lova A. An assessment of railway capacity. Transportation Research Part E. 2008;**44**:774-806

[5] Valentinovič L, Sivilevičius H. Railway line capacity methods analysis and their application in "Lithuanian Railways" justification. The 9th International Conference "Environmental Engineering". Vilnius, Lithuania; 2014

[6] Melody KS. Railway track capacity: Measuring and managing. [Doctoral thesis]. University of Southampton, Faculty of Engineering and the Environment; 2012

[7] Krmac E, Djordjevic B. An evaluation of train control information systems for sustainable railway using the analytic hierarchy process (AHP) model. European Transport Research Review. 2017:9-35. DOI: 10.1007/s12544-017-0253-9

[8] Bevrani B. Capacity determination and expansion models for rail networks. QUT Thesis (Masters by Research), Queensland University of Technology; 2015

[9] Pouryousef H. Timetable management technique in railway capacity analysis: Development of the hybrid optimization of train schedules (HOTS) model. [Dissertation]. Michigan Technological University; 2015

[10] Pachl J. Railway Operation and Control. USA: Mountlake Terrace-WA VTD Rail Publishing; 2002

[11] Sameni MK, Dingler M, Presto JM, Barkan CP. Profit-Generating Capacity for a Freight Railroad. Transportation Research Board (TRB) 90th Annual Meeting. Washington, DC; 2011

[12] Banverket. Application of the UIC Capacity Leaflet at Banverket. Banverket: Swedish National Rail Administration; 2005

[13] Kontaxi E, Ricci S. Railway Capacity Analysis: Methodological Framework and Harmonization Perspectives. 12th World Conference on Transport Research-WCTR. Lisbon, Portugal; 2010

[14] Burdett RL, Kozan E. Absolute capacity determination and timetabling in railways. Fifth Asia Pacific Industrial Engineering and Management Systems Conference, APIEMS; 2004

[15] Kozan E, Burdett R. A railway capacity determination model and rail access charging methodologies. Transportation Planning and Technology. 2005;**28**(1):27-45

[16] Burdett RL, Kozan E. Techniques for absolute capacity determination in railways. Transportation Research Part B. 2006;**40**:616-632

[17] Yuan J, Hansen AI. Optimizing capacity utilization of stations by estimating knock-on train delays. Transportation Research Part B. 2007;**41**:202-217

[18] Weik N, Niebel N, Nießen N. Capacity analysis of railway lines in Germany - A rigorous discussion of the queueing based approach. Journal of Rail Transport Planning and Management. 2016;**9**:99-115

[19] Mussone L, Wolfler RC. An analytical approach to calculate the capacity of a railway system. European Journal of Operational Research. 2013;**228**(1):11-23

[20] Wahlborg M. Banverket Experience of Capacity Calculations According to the UIC Capacity Leaflet. Computers in Railways, IX; 2004. pp. 665-673

[21] UIC Code 406. Capacity. International Union of Railways; 2004

[22] Lindner T. Applicability of the analytical UIC Code 406 compression method for evaluating line and station capacity. Journal of Rail Transport Planning and Management. 2011;**1**:49-57

[23] Lai YC, Liu, YH, Lin YJ. Development of base train equivalents for headway-based analytical railway capacity analysis. 5th International Seminar on Railway Operations Modelling and Analysis. Copenhagen, Denmark; 2013

[24] Lai Y-C, Liu Y-H, Lin Y-J. Standardization of capacity unit for headway-based rail capacity analysis. Transportation Research Part C. 2015;**57**:68-84

[25] Armstrong J, Preston J. Extending capacity utilisation measures from railway network links to nodes. WIT Transactions on The Built Environment. 2012;**127**

[26] Wittrup JL, Landex A, Anker NO. Assessment of stochastic capacity consumption in railway networks. 6th International Seminar on Railway Operations Modelling and Analysis. Tokyo, Japan; 2015

[27] Ljubaj I, Mlinarić TJ, Radonjić D. Proposed solutions for increasing the capacity of the Mediterranean Corridor on section Zagreb - Rijeka. Procedia Engineering. 2017;**192**: 545-550

[28] Jensen LW, Landex A, Nielsen OA, Kroon LG, Schmidt M. Strategic assessment of capacity consumption in railway networks: Framework and model. Transportation Research Part C. 2017;**74**:126-149

[29] De Kor AF, Heidergott B, Ayhan HAA. probabilistic approach for determining railway infrastructure capacity. European Journal of Operational Research. 2003;**148**:644-661

[30] Landex A, Kaas AH. Planning the most suitable travel speed for high frequency railway lines. 1st International Seminar on Railway Operations Modelling and Analysis. TU Delft; 2005

[31] Yaghini M, Sarmadi M, Nikoo N, Momeni M. Capacity consumption analysis using heuristic solution method for under construction railway routes. Networks and Spatial Economics. 2014a;**14**:317-333

[32] Yaghini M, Nikoo N, Ahadi HR. An integer programming model for analysing impacts of different train types on railway line capacity. Transport. 2014;**29**(1):28-35

[33] Zhang J. Analysis on line capacity usage for China high speed railway with optimization approach. Transportation Research Part A. 2015;**77**:336-349

[34] Burdett RL. Optimisation models for expanding a railway's theoretical capacity. European Journal of Operational Research. 2016;**251**:783-797

[35] Guo B, Zhou L, Yue Y, Tan J. A study on the practical carrying capacity of large high-speed railway stations considering train set utilization. Mathematical Problems in Engineering. 2016:11

[36] Ortega Riejos FA, Barrena E, Ortiz JD, Laporte G. Analyzing the theoretical capacity of railway networks with a radial-backbone topology. Transportation Research Part A. 2016;**84**:83-92

[37] Pellegrini P, Marliere G, Rodriguez J. RECIFE-SAT: A MILP-based algorithm for the railway saturation problem. Journal of Rail Transport Planning and Management. 2017; **7**:19-32

[38] Landex A. Methods to estimate railway capacity and passenger delays [thesis]. Technical University of Denmark; 2008

[39] Kaas AH. Strategic Capacity Analysis of Networks: Developing and Practical use of Capacity Model for Railway Networks. ScanRail Consult, Technical University of Denmark; 1991

[40] Han S, Yue Y, Zhou L. Carrying capacity of railway station by microscopic simulation method. IEEE 17th International Conference on Intelligent Transportation Systems (ITSC). Qingdao, China; 2014

[41] Malavasi G, Molková T, Ricci S, Rotoli F. A synthetic approach to the evaluation of the carrying capacity of complex railway nodes. Journal of Rail Transport Planning and Management. 2014;**4**:28-42

[42] Lai Y, Barkan C. Enhanced parametric railway capacity evaluation tool. Journal of the Transportation Research Board. 2009;**2117**:33-40

[43] Yung-Cheng L, Yung-An H. Estimation of single and double-track capacity with parametric models. Transportation Research Board 91st Annual Meeting; 2012

[44] Sogin SL, Lai Y-CR, Dick TC, Barkan CP. Comparison of capacity of single- and double-track rail lines. Transportation Research Record: Journal of the Transportation Research Board. 2013;**2374**:111-118

[45] Lai Y-CR, Liu Y-H, Lin T-Y. The development of base train equivalents to standardize trains for capacity analisys. Transportation Research Record. 2012:119-125

[46] Mattalia A. The Effects on Operation and Capacity on Railways Deriving from the Switching to Continuous Signals and Tracing Systems (ERTMS). Stockholm: KTH Royal Institute of Technology; 2007

[47] Magnarini M. Evaluation of ETCS on railway capacity in congested areas – A case study within the network of Stockholm [Master thesis]. Stockholm: KTH Royal Institute of Technology; 2010

[48] McNaughton A. Signalling Headways and Maximum Operational Capacity on High Speed Two London to West Midlands Route; 2011

[49] Lai Y-C, Wang S-H. Development of analytical capacity models for conventional railways with advanced signaling systems. Journal of Transportation Engineering. 2012;**138**(7):961-974

[50] Goverde RM, Corman F, D'Ariano A. Railway line capacity consumption of different railway signalling systems under scheduled and disturbed conditions. Journal of Rail Transport Planning and Management. 2013;**3**:78-94

[51] Coviello N. The influence of ETCS and traffic composition on daily capacity of single track lines: An evaluation model developed and applied on the trans-Mongolian railway. [Master thesis]. Stockholm, Sweden: Royal Institute of Technology; 2013

[52] Li W, Tang T. The impact of signaling system heterogeneity on urban transit capacity. 2013 IEEE International Conference on Intelligent Rail Transportation Proceedings. Beijing, China

[53] Quaglietta E. A simulation-based approach for the optimal design of signalling block layout in railway networks. Simulation Modelling Practice and Theory. 2014;**46**:4-24

[54] Venkateswaran KG, Nicholson GL, Roberts C, Stone R. Impact of automation on the capacity of a mainline railway: A preliminary hypothesis and methodology. 18th International Conference on Intelligent Transportation Systems; 2015

[55] Kontaxi E, Ricci S. Techniques and methodologies for railway capacity analysis: Comparative studies and integration perspectives. 5th International Scientific Conference "Theoretical and Practical Issues in Transport". Pardubice; 2010

[56] Kontaxi E, Ricci S. Railway capacity handbook: A systematic approach to methodologies. Procedia - Social and Behavioral Sciences. 2012;**48**:2689-2696

[57] Palumbo M. Railway Signalling since the birth to ERTMS. railwaysignalling.eu; 2013

[58] Abed SK. European rail traffic management system – An overview. International Conference on Energy, Power and Control (EPC-IQ). Basrah, Iraq. 2010

[59] Ghazel M. Formalizing a subset of ERTMS/ETCS specifications for verification purposes. Transportation Research Part C. 2014;**42**:60-75

[60] Franklin F, Nemtanu F, Teixeira PF. Rail infrastructure, ITS and access charges. Research in Transportation Economics. 2013;**41**:31-42

[61] Smith P, Majumdar A, Ochieng WY. An overview of lessons learnt from ERTMS implementation in European railways. Journal of Rail Transport Planning and Management. 2012;**2**:79-87

[62] Parent de Curzon E. ERTMS - a contribution to the creation of tomorrow's railway. Elektrotechnik und Informationstechnik. 1999;**116**(9):499-503

[63] Collins L. Mixed Signals: A common European rail signalling system could save lives and money. IEEE Review; 2005

[64] Durmus MS, Yildirim U, Soylemez MT. Interlocking system design for ERTMS/ETCS: Approach with batches petri nets. IFAC Proceedings. 2012;**45**:110-115

[65] Th AU. Report on the Elimination of Level Crossing and on Renewal of Existing ICT Devices. ACROSSEE Project; 2014

[66] Sameni MK, Landex A. Capacity utilization in European railways: Who is the fairest of them all? Transportation Research Board (TRB) 92nd Annual Meeting Transportation Research Board. Washington, DC; 2013

[67] Liu JS, Lu LY, Lu W-M, Lin JY. Data envelopment analysis 1978-2010: A citation-based literature survey. Omega 2012. DOI:10.1016/j.omega.2010.12.006

[68] Charnes A, Cooper WW, Rhode E. Measuring efficiency of decision making units. European Journal of Operational Research. 1978;**2**:429-444. DOI: 10.1016/0377-2217(78)90138-8

[69] Liu WB, Meng W, Li XX, Zhang DQ. DEA models with undesirable inputs and outputs. Annals of Operations Research. 2010;**173**:177-194

[70] Banker RD, Charnes A, Cooper WW. Some models for estimating technical and scale inefficiencies in data envelopment analysis. Management Science. 1984;**30**:1078-1092. DOI: 10.1287/mnsc.30.9.1078

[71] Charnes A, Cooper WW, Golany B, Seiford L. Foundations of data envelopment analysis for Pareto-Koopmans efficient empirical production functions. Journal of Econometrics. 1985;**30**:91-107

[72] Charnes A, Cooper WW, Wei QL, Huhng Z. Cone ratio data envelopment analysis and multi-objective programming. International Journal of Systems Science. 1989;**20**: 1099-1118

[73] Meng F, Su B, Thomson E, Zhou D, Zhou P. Measuring China's regional energy and carbon emission efficiency with DEA models: A survey. Applied Energy. 2016;**183**:1-21

[74] Wu J, Zhu Q, Yin P, Song M. Measuring energy and environmental performance for regions in China by using DEA-based Malmquist indices. Operational Research International Journal. 2015. DOI: 10.1007/s12351-015-0203-z

[75] Farrell JM. The measurement of productive efficiency. Journal of the Royal Statistical Society A. 1957;**120**:253-281

[76] Zhou P, Ang BW, Poh KL. A survey of data envelopment analysis in energy and environmental studies. European Journal of Operational Research. 2008;**189**:1-18. DOI: 10.1016/j.ejor.2007.04.042

[77] Cooper WW, Seiford LM, Tone K. Introduction to Data Envelopment Analysis and its Use: With DEA-Solver Softwer and References. New York: Springer; 2006

[78] Liu W, Zhou Z, Ma C, Liu D, Shen W. Two-stage DEA models with undesirable input-intermediate-outputs. Omega. 2015;**56**:74-87

[79] Vencheh AH, Matin RK, Kajani MT. Undesirable factors in efficiency measurement. Applied Mathematics and Computation. 2005;**163**:547-552

[80] Amirteimoori A, Kordrostami S, Sarparast M. Modeling undesirable factors in data envelopment analysis. Applied Mathematics and Computation. 2006;**180**:444-452

[81] Seiford LM, Zhu J. Modeling undesirable factors in efficiency evaluation. European Journal of Operational Research. 2002;**142**:16-20

[82] Jahanshahloo GR, Vencheh AH, Foroughi A. Inputs/outputs estimation in DEA when some factors are undesirable. Applied Mathematics and Computation. 2004;**156**:19-32

[83] Jahanshahloo GR, Lotfi FH, Shoja N, Tohidi RG. Undesirable inputs and outputs in DEA models. Applied Mathematics and Computation. 2005;**169**:917-925

[84] Zhou P, Ang BW. Linear programming models for measuring economy-wide energy efficiency performance. Energy Policy. 2008;**36**:2911-2916. DOI: 10.1016/j.enpol.2008.03.041

[85] Zhou G, Chung W, Zhang Y. Measuring energy efficiency performance of China's transport sector: A data envelopment analysis approach. Expert Systems with Applications. 2014;**41**:709-722. DOI: 10.1016/j.eswa.2013.07.095

[86] Bian Y, Jang F. Resource and environment efficiency analysis of provinces in China: A DEA approach based on Shannon's entropy. Energy Policy. 2010;**38**(4):1909-1917. DOI: 10.1016/j.enpol.2009.11.071

[87] Wang K, Lu B, Wei YM. China's regional energy and environmental efficiency: A range-adjusted measure based analysis. Applied Energy. 2013;**112**:1403-1415

[88] Boysen HE. General model of railway transportation capacity. WIT Transactions on The Built Environment. 2012;**127**

[89] OBB. OBB in numbers: We move Austria forward; 2016

[90] Cooper WW, Li S, Seiford ML, Tone K, Trall MR, Zhu J. Sensitivity and stability analysis in DEA: Some recent developments. Journal of Productivity Analysis. 2001;**15**:217-246

[91] Gandibleux X, Pierre R, Xavier D. RECIFE: A MCDSS for railway capacity evaluation. Multiple criteria decision making for sustainable energy and transportation systems. In: Proceedings of the 19th International Conference on Multiple Criteria Decision Making, Auckland, New Zealand, January 7-12. 2008. pp. 93-103

[92] Feretti V, Degioanni A. How to support the design and evaluation of redevelopment projects for disused railways? A methodological proposal and key lessons learned. Transportation Research Part D. 2017;**52**:29-48

[93] Yinshang W, Yi G, Dingwen D, Dewen L. Public places safety management evaluation of railway stations. Procedia Engineering. 2012;**45**:240-247

[94] Chen X, Chang Y. The study of evacuation passenger service level of Shanghai-Nanjing high-speed railway stations. Procedia - Social and Behavioral Sciences. 2013;**96**:265-269

[95] Bureika G, Bekintis G, Liudvinavičius L, Vaičiunas G. Applying analytic hierarchy process to assess traffic safety risk of railway infrastructure. Eksploatacja i Niezawodnosc – Maintenance and Reliability. 2013;**15**(4):376-383

[96] Polat G, Eray E. An integrated approach using AHP-ER to supplier selection in railway projects. Procedia Engineering. 2015a;**123**:415-422

[97] Bersani C, Guerisoli C, Mazzino N, Sacile R, Sallak M. A multi-criteria methodology to evaluate the optimal location of a multifunctional railway portal on the railway network. Journal of Rail Transport Planning and Management. 2015;**5**:78-91

[98] Dhir S, Marinov MV, Worsley D. Application of the analytic hierarchy process to identify the most suitable manufacturer of rail vehicles for High Speed 2. Case Studies on Transport Policy. 2015;**3**:431-448

[99] Gercek H, Karpak B, Kilincaslan T. A multiple criteria approach for the evaluation of the rail transit networks in Istanbul. Transportation. 2004;**31**:203-228

[100] Sivilevičius H, Maskeliūnaitė L. The criteria for identifying the quality of passengers` transportation by railway and their ranking using AHP method. Transport. 2010; **25**(4):368-381

[101] Maskeliūnaitė L, Sivilevičius H, Podvezko V. Research on the quality of passenger transportation by railway. Transport. 2009;**24**(2):100-112

[102] Maskeliunaite L, Sivilevicius H. Applying AHP Technique to the Assessment of Railway Trip Quality (RTQ). 8th International Conference Environmental Engineering. Vilnius, Lithuania; 2011

[103] Nystrom B, Soderholm P. Selection of maintenance actions using the analytic hierarchy process (AHP): Decision-making in railway infrastructure. Structure and Infrastructure Engineering. 2010;**6**(4):467-479

[104] Yuanyuan S, Gang Z, Zhangdui Z. Interference assessment in the high-speed railway wireless communication networks by AHP-GRAP. Fifth Conference on Measuring Technology and Mechatronics Automation; 2013

[105] Sivileviãius H, Maskelinaitò L. The Numerical Example for Evaluating the Criteria Describing the Quality of the Trip by International Train. Ekonomika a Management. 2014;**7**(2):74-86

[106] Shapiro AF, Koissi M-C. Fuzzy logic modifications of the Analytic Hierarchy Process. Insurance: Mathematics and Economics. 2017;**75**:189-202

[107] Li Y-H. Design scheme selection of railway freight car using a fuzzy analytic hierarchy process. Sixth International Conference on Fuzzy Systems and Knowledge Discovery; 2009

[108] An M, Chen Y, Baker CJ. A fuzzy reasoning and fuzzy-analytical hierarchy process based approach to the process of railway risk information: A railway risk management system. Information Sciences. 2011;**181**:3946-3966

[109] Aydin N, Celik E, Gumus AT. A hierarchical customer satisfaction framework for evaluating rail transit systems of Istanbul. Transportation Research Part A. 2015;**77**:61-81

[110] Jiang P, Wang D, Xing Y. Risk analysis of general accidents in china railway passenger transportation. Seventh International Conference on Measuring Technology and Mechatronics Automation. Nanchang, China; 2015

[111] Grozdanovic M, Janackovic GL, Stojiljkovic E. The selection of the key ergonomic indicators influencing work efficiency in railway control rooms. Transactions of the Institute of Measurement and Control. 2016;**30**(10):1174-1185

[112] Cheng Y-H, Tsao H-L. Rolling stock maintenance strategy selection, spares parts'estimation, and replacements' interval calculation. International Journal of Production Economics. 2010;**128**:404-412

[113] Aragonés-Beltrán P, García-Melón M, Montesinos-Valera J. How to assess stakeholders' influence in project management? A proposal based on the Analytic Network Process. International Journal of Project Management. 2017;**35**:451-462

[114] Chang Y-H, Wey W-M, Tseng H-Y. Using ANP priorities with goal programming for revitalization strategies in historic transport: A case study of the Alishan Forest Railway. Expert Systems with Applications. 2009;**36**:8682-8690

[115] Janic M. Multicriteria evaluation of high-speed rail, transrapid maglev and air passenger transport in Europe. Transportation Planning and Technology. 2003;**26**(6):491-512

[116] Chen S, Leng Y, Mao B, Liu S. Integrated weight-based multi-criteria evaluation on transfer in large transport terminals: A case study of the Beijing South Railway Station. Transportation Research Part A. 2014;**66**:13-26

[117] Jinbao R, Xing Z. Fault Criticality Evaluation of Metro Door based on modified FMEA. Proceedings of the 33rd Chinese Control Conference. Nanjing, China; 2014

[118] Aydin N. A fuzzy-based multi-dimensional and multi-period service quality evaluation outline for rail transit systems. Transport Policy. 2017;**55**:87-98

[119] Rotoli F, Cawood EN, Christidis P. A data envelopment snalysis approach for accessibility measures: Simulating operational enhancement scenarios for railway across Europe. European Transport Research Review. 2015;**7**(18)

[120] Martin JC, Reggiani A. Recent methodological developments to measure spatial interaction: Synthetic accessibility indices applied to high-speed train investments. Transport Reviews. 2007;**27**(5):551-571

[121] Merkert R, Smith AS, Nash CA. Benchmarking of train operating firms - a transaction cost efficiency analysis. Transportation Planning and Technology. 2010;**33**(1):35-53

[122] Zhang Y, Lei D, Li X, Fu Y. The analysis of shunting locomotives' operating efficiency based on gray-DEA. Research Journal of Applied Sciences, Engineering and Technology. 2013;**5**(5):1720-1725

[123] Li Z. Operation performance evaluation and optimization based on SUPER-SBM DEA model in railway industry in China. International Conference on Information Science and Cloud Computing Companion. Guangzhou, China; 2013

[124] Sama M, Meloni C, D'Ariano A, Corman F. A multi-criteria decision support methodology for real-time train scheduling. Journal of Rail Transport Planning and Management. 2015;**5**:146-162

[125] Sameni MK, Preston J, Sameni MK. Evaluating efficiency of passenger railway stations: A DEA approach. Research in Transportation Business & Management. 2016;**20**:33-38

[126] Kleinova E. Does liberalization of the railway industry lead to higher technical effectiveness? Journal of Rail Transport Planning and Management. 2016;**6**:67-76

[127] Cavone G, Dotoli M, Epicoco N, Seatzu C. A decision making procedure for robust train rescheduling based on mixed integer linear programming and Data Envelopment Analysis. Applied Mathematical Modelling. 2017;**52**:255-273

[128] Kosijer M, Ivić M, Marković M, Belošević I. Multicriteria decision-making in railway route planning and design. Građevinar. 2012;**64**(3):195-205

[129] Celik E, Aydin N, Gumus AT. A multiattribute customer satisfaction evaluation approach for rail transit network: A real case study for Istanbul, Turkey. Transport Policy. 2014;**36**:283-293

[130] Yong F, Yong Q, Shuting Z, Wantong L, Limin J, Xinwang L, et al. Operation safety assessment of high-speed train with fuzzy group decision making method and empirical research. 2nd International Conference on Cloud Computing and Internet of Things, Dalian, China; 2016

[131] Saat MR, Serrano JA. Multicriteria high-speed rail route selection: Application to Malaysia's high-speed rail corridor prioritization. Transportation Planning and Technology. 2015;**38**(2):200-213

[132] Azadeh A, Ghaderi SF, Izadbakhsh H. Integration of DEA and AHP with computer simulation for railway system improvement and optimization. Applied Mathematics and Computation. 2008;**195**:775-785

[133] Mohajeri N, Amin GR. Railway station site selection using analytical hierarchy process and data envelopment analysis. Computers & Industrial Engineering. 2010;**59**:107-114

[134] Jafarian-Moghaddam AR, Ghoseiri K. Fuzzy dynamic multi-objective data envelopment analysis model. Expert Systems with Applications. 2011;**38**:850-855

[135] Mandic D, Jovanovic P, Bugarinovic M. Two-phase model for multi-criteria project ranking: Serbian Railways case study. Transport Policy. 2014;**36**:88-104

[136] Hu X, Li X, Huang Y. Urban rail transit risk evaluation with incomplete information. Procedia Engineering. 2016;**137**:467-477

[137] Ozceylan E, Erbas M, Tolon M, Kabak M, Durgut T. Evaluation of freight villages: A GIS-based multi-criteria decision analysis. Computers in Industry. 2016;**76**:38-52

[138] Nazeri A, Naderikia R. A new fuzzy approach to identify the critical risk factors in maintenance management. The International Journal of Advanced Manufacturing Technology. 2017;**92**:3749-3783

[139] Zhan Q, Zheng W, Zhao B. A hybrid human and organizational analysis method for railway accidents based on HFACS-Railway Accidents (HFACS-RAs). Safety Science. 2017;**91**:232-250

4

Improving Feasibility of High-Speed Train Project: Creating Added Value

Mohammed Ali Berawi

Abstract

Infrastructure plays a significant role in increasing economic development by providing access of transportation and improving connectivity. High-speed train (HST), one of mega infrastructure projects, has a positive impact on economic development of a nation. However, the project feasibility requires the maximum value for money and an acceptable risk to attract private investors. This study aims to improve the feasibility of the project by producing a conceptual design of Jakarta-Surabaya high-speed train in Indonesia. Value engineering will be used to evaluate both technical and financial aspects of the project. The methodology uses both qualitative and quantitative approaches through a case study, in-depth interviews, and life-cycle cost analysis. The result shows an optimum route sketching for the project and potential added value to the project. It consists of the solar cell, fiber optic, tourism, and transit-oriented development. The output also generates the division of responsibility between the government and business entity during the project lifecycle regarding the project financing. The institutional scheme will regulate the position and roles for each related stakeholder that was involved in the HST project development.

Keywords: infrastructure, feasibility, high-speed train, mega project, value engineering

1. Introduction

Infrastructure development is one of the most significant aspects in accelerating economic and social growth [1]. Global competitiveness report published that the quality and quantity of infrastructure improves the ease of investment and creates more productive activities. Access for the people to use infrastructure is the key to accommodate individual activities and community engagement [2]. Reducing the distance among regions and establishing connectivity of national and international markets are significant to improve economic growth of a nation [3]. Infrastructure itself contributed about 60% to the economic growth of the United States [4]. Many other countries also experience similar benefits from the infrastructure and transportation connectivity to the people's daily activity and mobility.

Indonesia's gross domestic product (GDP) mainly depends on the economic development in Java Island. Jakarta located in western part of the island contributes about 20%, while Surabaya in the opposite location produces 6% of the GDP [5]. Location of both cities in the national context can be seen in **Figure 1**. The distance between those cities is approximately about 700 km and can be accessed through particular transportation modes with different travel times. Private vehicles require 22–26 h, 10–14 h by train and 1.5 h by airplane. Travel by plane saves the time but requires a double-to-triple cost compared to other transportation modes. Private vehicles consume less amount of cost but needs longer travel time to reach the destination. On the other hand, rail transportation offers competitive price compared to air transportation but proposes faster travel time over private vehicles and in some way, a minimum gap of travel time to air transportation [6].

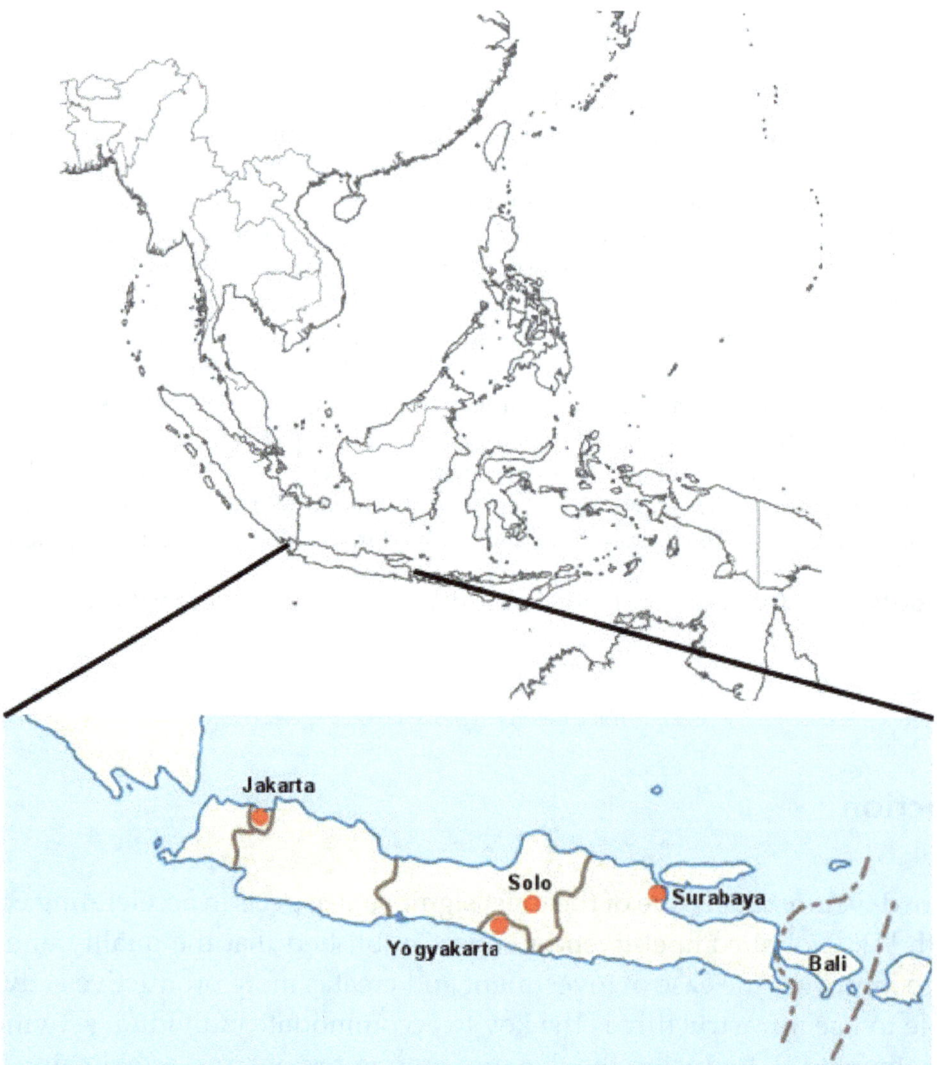

Figure 1. City of Jakarta and Surabaya in Indonesia context.

In 2012, the Government of Indonesia launched a high-speed train (HST) project to connect Jakarta and Surabaya. The project has been offered to national and international investors, but progress of the project remains limited due to technical issues and financial deficiency. The needs of massive investment become the main issues investors tend to hold their investment, and the government has insufficient funds to support the project.

Levinson [7] argued that high capital costs and passengers' demand play a significant role to attract business entities. The United States requires 6–9 million passengers per year to meet the targeted return on investment, and thus the HST project in the country remains on hold. A high-occupancy high-speed train only reaches a maximum of 3.4 million passengers per year. Based on the situation, the study uses an alternative approach through value engineering (VE) process to improve the feasibility of the HST project. It has been proven to produce a strategic outcome regarding quality [8, 9], technology breakthrough [10], efficiency [11], and innovative creation [12]. VE generates value by proposing more benefits over cost. Transportation project in the United States experiences cost saving for VE implementation about 5.9% from the total cost of the project in 2015.

The study advocates the alternative use of value engineering to improve the performance of the infrastructure project. The result expected can be used for practical implementation in the industry, assist the decision-making process and regulatory framework for government institutions, as well as knowledge dissemination and debates for academic purposes.

2. Jakarta-Surabaya route planning

The route for high-speed train considers technical issues such as topography, length of the route, population, and economic development in a region expressed by gross regional domestic product (GRDP). Cities in western part of Java Island are mostly located at high elevations with mountains and cliffs. The trajectory recommends passing flat surface and minimizes the use of bridge and tunnel to reduce higher costs. From the analysis, green areas are recommended for the high-speed train line due to the flat surface. Moreover, the yellow to the orange area is chosen when the slope and topography meet the minimum technical requirement. The elevation map of Java can be seen in **Figure 2**.

The analysis identifies potential 9 cities out of 67 cities in Java Island for the high-speed train station. Jakarta ranked as the highest contributor in terms of GRDP and population. It is followed by Surabaya, Yogyakarta, Bandung, and Semarang. Four cities that are categorized into a mid-size city in terms of the population also included Kediri, Cilacap, Cirebon, and Solo due to their enormous contribution to the regional economic development. The location for each city can be seen in **Figure 3**.

The nine cities generate seven alternative routes, and each of them has a varied length from 754 to 958.6 km. Further analysis identified two possible routes for the project. Route 1 connects the city of Jakarta-Cilacap-Yogyakarta-Semarang-Solo-Kediri and Surabaya. The total distance for this route is about 958.6 km. Route 1 can be seen in **Figure 4**.

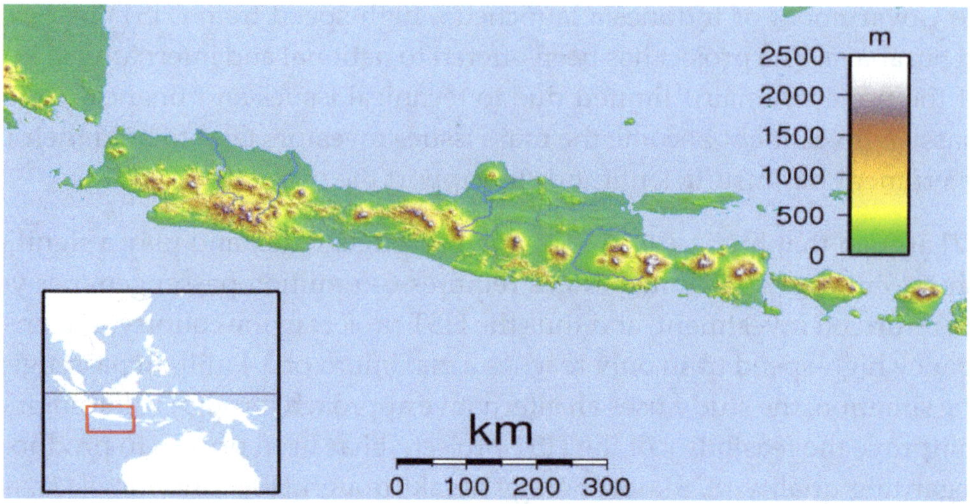

Figure 2. Elevation map of Java Island.

Figure 3. Nine potential cities for high-speed train station.

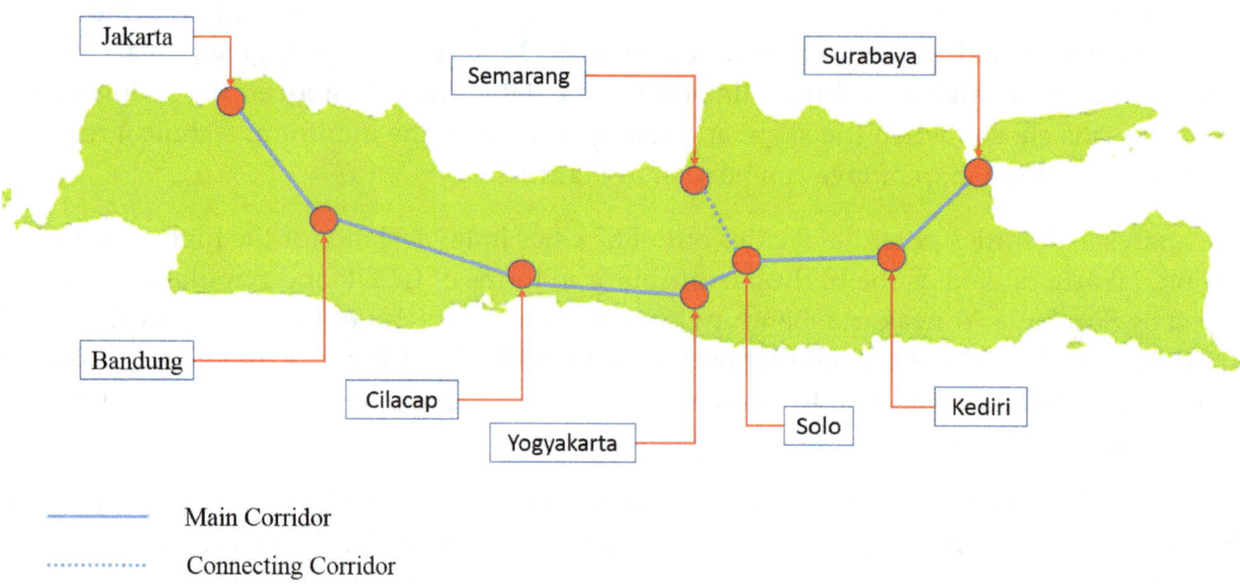

Figure 4. Route 1 from Jakarta – Surabaya high-speed train.

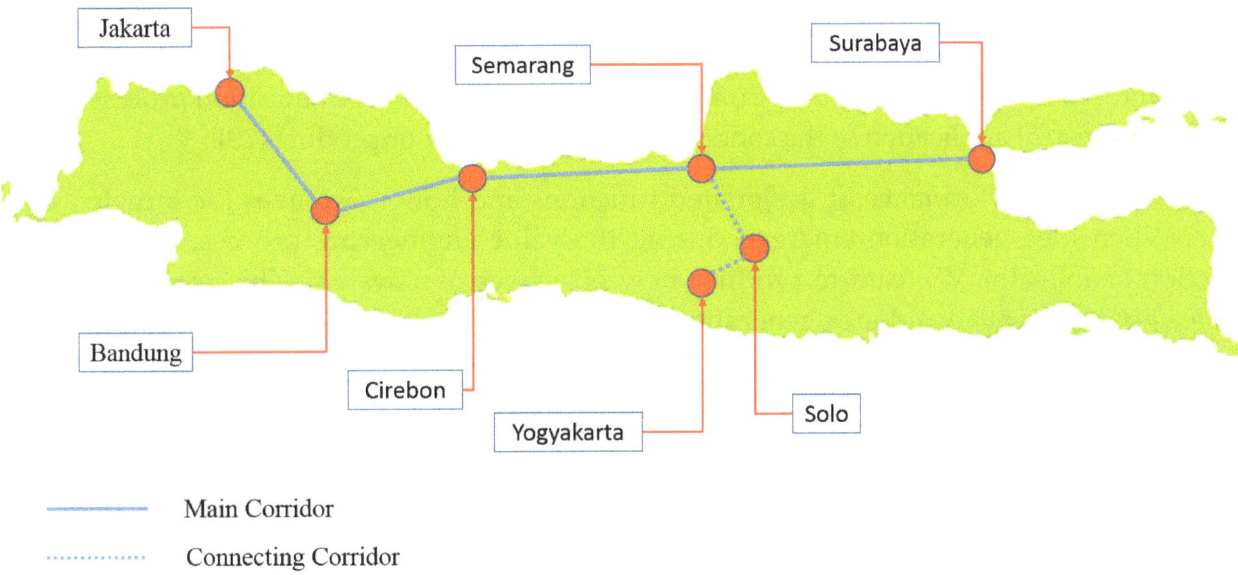

Figure 5. Route 2 from Jakarta-Surabaya high-speed train.

On the other hand, route 2 connects Jakarta-Bandung-Cirebon-Semarang (+Solo-Yogyakarta)-Surabaya. The total distance for this path is about 868.5 km, as shown in **Figure 5**. Each section in both alternative routes has different lengths and varied use of infrastructure components (rail structure, viaduct, bridge, or tunnel). The components were determined by estimating the topography of sections using Google Earth program. The maximum slope also affects the speed of the train [13]. As topography contributes to the cost of construction, careful planning must further be evaluated.

3. Value engineering of Jakarta-Surabaya HST project

Value engineering (VE) is a multidisciplinary approach that analyses and systematically improves function by reducing the cost and increasing the value of a product, design, system, or service.

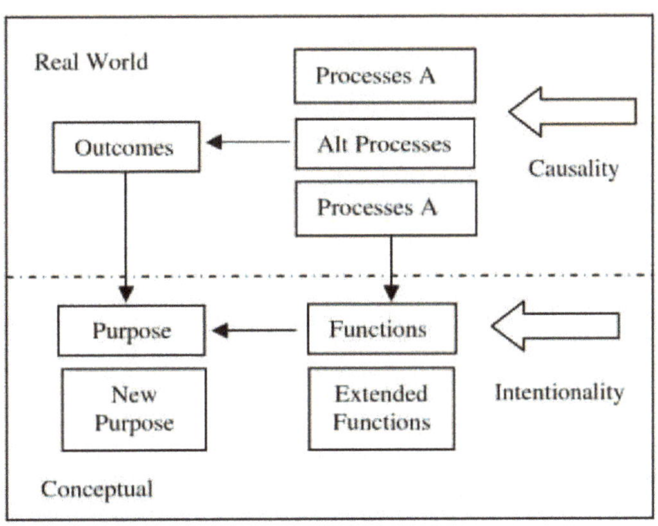

Figure 6. Functional thinking enables innovation.

VE is argued as a proven method to solve problems and to improve projects' competitiveness [14]. VE capability to increase competitiveness driven by the ability to produce creative ideas and innovation [15–17]. Innovation made from the creativity stage [18], formalization processes [19], and the successful application of the concept concerning output or product [20].

It measures the performance of designated functions and how it achieves the targeted purpose. When idea generation emerged during the value engineering process, the extended function enables the VE team to produce new objectives of a system. This process demonstrates how innovative and idea generation might improve the whole perspective in conceptual thinking and contribute to the real world [8]. The concept can be seen in **Figure 6**.

The Jakarta-Surabaya HST project development adopts value engineering job plan [8]. Firstly, information phase was conducted by observing literature study about the project detail. Secondly, function analysis is developed by identifying the function from existing data and information as follows:

- Scope of the problem under study: high-speed train between Jakarta and Surabaya
- Highest order function: stimulate economic growth
- Lowest order function: generate income
- Design objective: create added value and develop infrastructure connectivity
- Basic function: increasing mobility
- Dependent functions: transport people and goods
- Processes: construct line

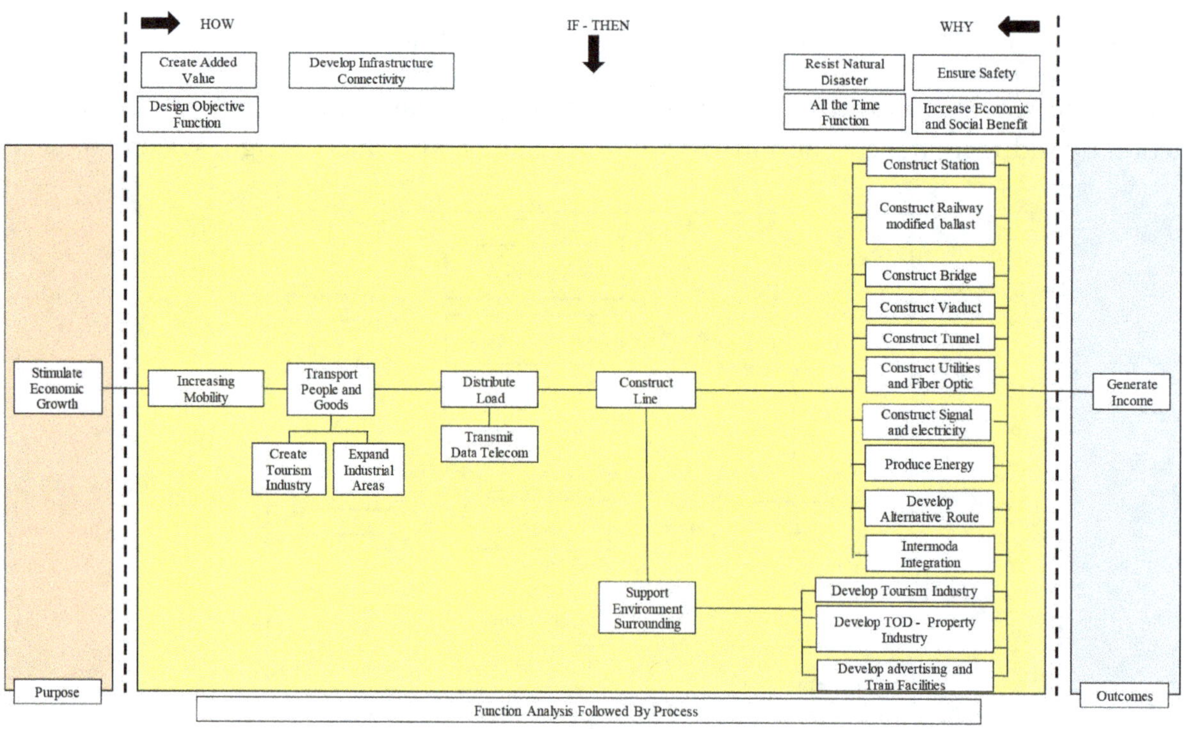

Figure 7. FAST diagram for Jakarta-Surabaya high-speed train.

The logical order of the previous function is mapping through a FAST diagram tool. The 'how-why' logical model is used to identify, classify, develop, and select functions that could create higher value and benefit to the project development. The FAST diagram, as shown in **Figure 7**, produces four additional functions such as solar cell, tourism, transit-oriented development (TOD), and fiber optic. Evaluation phase will be used to assess the benefits of additional functions from the financial perspective. Furthermore, public-private partnership (PPP) acts as the development phase to investigate the identified functions and financial engineering. It attempts to find the best formulation for both private investors and government in financing the project.

4. Financial scenario

Financial scenario elaborates initial cost, operational and maintenance cost, and revenue to generate an internal rate of return (IRR) and net present value (NPV). Initial cost, operation, and maintenance cost of high-speed train infrastructure consider railway infrastructure such as tracks, signaling, electricity, tunnel, bridge, and station area.

The scenario for revenue uses three categories of income from the low tariff, mid-tariff, and high tariff. Several assumptions are proposed for estimation boundary that is retrieved from Statistics Indonesia and other reliable sources. The discount rate is set about 6.81%. Inflation categorized from general inflation, transportation, fossil fuel, electricity, and telecommunication to property ranges from 1.63 to 5.95%. The scenario will compare the single function and multifunction from the life-cycle cost perspective. The result shows the benefits from value engineering process by an increased value of the rate of return.

4.1. Single function

The analysis produces route two from Jakarta-Bandung-Cirebon-Semarang-(Solo-Yogyakarta)-Surabaya as the main path of the high-speed train. The initial cost of the route is about 118.64 trillion rupiahs for 868.5 km or equal to 136.60 billion rupiahs/km. The operational and maintenance cost projected is about 5.3 trillion rupiahs/year or equal to 6.10 billion rupiahs/year. Revenue of HST considers demand forecast based on the shifting from the existing transportation mode such as the airplane, private vehicles, and regular train. The tariff for HST of Jakarta–Surabaya is set into three scenarios from 600,000 rupiahs, 750,000 rupiahs, and 1,000,000 rupiahs.

The result shows that low tariff produces a 5.12% of the internal rate of return (IRR), while mid-tariff claimed 8.7% of IRR and 13.8% for the high tariff. The low and mid-tariffs are failed to reach a minimum attractive rate of return (MARR) of the most infrastructure investors approximately about 12%. Subsequently, further elaboration is required by proposing additional functions to increase the feasibility of the project.

4.2. Multifunction

In a multifunction concept, the additional function includes solar cell, fiber optic, tourism, and transit-oriented development (TOD). Solar cell's initial cost depends on the number of solar cells and its capacity. Nine units of solar cells were generated with each one's capacity as about 70 kW with a unit efficiency of 20.83%. Overall, the initial cost is about 32.02 billion rupiahs.

Typology	Location	Description
Type 1	Jakarta, Surabaya, and Bandung	Land area: 10,000 sqm Residence area: 61,000 sqm Offices area: 19,000 sqm Commercial area : 3700 sqm Parking area : 1400 sqm
Type 2	Cirebon, Yogyakarta, and Semarang	Land area : 9000 sqm Residence area : 41,000 sqm Offices area: 13,000 sqm Commercial area : 2500 sqm Parking area : 1200 sqm
Type 3	Solo, Cilacap, Kediri	Land area : 8000 sqm Residence area : 31,000 sqm Offices area: 10,000 sqm Commercial area : 2000 sqm Parking area : 700 sqm

Table 1. Transit-oriented development typology.

The operation and maintenance cost is 1% of the initial cost or equal to 25.09 billion rupiahs/year and is assumed to increase about 5.9% per year. Revenue considers baseline tariff from the government for about 1000 rupiahs and is multiplied by the electricity output from the solar cell. It will increase 10% per 2 years. Revenue for the low tariff is about 129.57 billion rupiahs; mid-tariff equals to 194.35 billion rupiahs; and high tariff around 259.13 billion rupiahs.

Fiber optic considers material cost and installation cost to determine its investment. The initial cost is about 571.73 billion rupiahs and 678.99 billion rupiahs/year for operational and maintenance cost. Revenue is generated from the number of services per month that charged about 2 million rupiahs. The result shows the lowest to highest annual revenue of fiber optic by 9.53 trillion rupiahs, 11.91 trillion rupiahs, and 13.11 trillion rupiahs, respectively.

The initial cost of tourism follows a benchmark from other tourism area developmental sectors and considers global cost construction index. The initial cost is estimated as about 17.31 trillion rupiahs. Operational and maintenance cost is 23.22 trillion rupiahs. Revenue is generated from ticketing and other related components such as retails, parking area, and commercials. Ticket proposes for about 300,000 rupiahs per person and increases 10% per two years. The revenue ranges from 83.94 trillion rupiahs, 119.91 trillion rupiahs, and 155 trillion rupiahs.

Lastly, the initial cost of TOD follows several factors such as land acquisition, apartment, commercials, offices, and parking area. Each cost component is multiplied by land area for selected cities. The initial cost for TOD is about 5.24 trillion rupiahs. Operation and maintenance (O&M) cost only considers the parking area since apartment, commercials, and offices have been handled by the users and tenants. The O&M cost is estimated for about 276.17 billion rupiahs. Revenue consists of apartment sales, offices rental, commercial leasing, and parking

Function	IC	OM	Revenue		
	(Rp. Billion)	(Rp. Billion)	Low (Rp. Billion)	Intermediate (Rp. Billion)	High (Rp. Billion)
HST	118,643.14	203,917.52	443,799.16	554,748.95	739,665.26
Commercial and Adv	821.83	236.91	5,799.12	10,902.36	16,005.57
TOD	5,244.67	276.17	125,475.38	159,961.23	179,291.61
Tourism	17,315.74	23,219.10	83,940.69	119,915.27	155,889.86
Fiber optic	571.73	678.99	9532.72	11,915.90	13,107.49
Solar cell	32.02	25.09	129.57	194.35	259.13
Total	142,629.13	228,353.78	668,676.64	857,638.06	1,104,218.92

Table 2. Life cycle cost analysis of Jakarta-Surabaya high-speed train.

usage. The TOD is divided into three types by considering the regional economy and population density. The location of each type depends on the size of cities, potential economic development, and land availability; thus, the revenue ranges from 138.70 trillion rupiahs, 178.49 trillion rupiahs, and 202.01 trillion rupiahs. The details of TOD typology can be seen in **Table 1**.

Overall, the life-cycle cost for the high-speed train, Jakarta-Surabaya, can be generated, as shown in **Table 2**. The total initial cost is about 142.63 trillion rupiahs, while operational and maintenance cost is about 228.35 trillion rupiahs and a range of revenue from 668.67 trillion rupiahs to 1,104.22 trillion rupiahs.

The result of the multifunction project shows an increased IRR compared to a single function of a high-speed train. Despite the increased feasibility, further evaluation through public-private partnership shall be conducted to elaborate the optimum financing scheme for government and investors.

5. Public-private partnership

Public-private partnership scheme in the project comprises scenarios by considering cost sharing from initial cost, operational and maintenance cost, as well as revenue between government and business entity. The cost sharing is conducted among functions because each of them generates a different internal rate of return (IRR), thus affecting investors' interest to involve in the project. The IRR should meet the investors' expectation by reaching their minimum attractive rate of return (MARR) and above interest rate.

Each function simulates 36 main scenarios and generates 252 scenarios in total to propose a maximum financial scheme output. Overall, initial cost and operation and maintenance cost consist of three assumptions from 40, 50, and 60%. It considers a similar division of responsibility between government and business entity as well as when one party should be responsible to the other. Sharing in revenue attempts to accommodate private interest by giving a higher percentage from 50 to 80%.

Most of the initial cost component in this project such as the high-speed train, fiber optic, tourism, and TOD use a share 40% government and 60% business entity. It is because the national budget plan is limited while the need for infrastructure financing is enormous; thus, transferring the cost to the business entity is most preferred. In terms of operational and maintenance cost, equal responsibility in the high-speed train, fiber optic and tourism between government and business entity are selected to minimize monopoly by the business. When the government has shared in this phase, the ticket price will be adjusted properly to match user ability to pay and their willingness to pay. Conversely, operation and maintenance of solar cell is relatively small compared to other functions. Thus, government portion in the service has a limited effect on the national budget plan. Lastly, the cost of TOD in this phase is only for the parking area; therefore, private entity proposes to be responsible for the cost.

The division of revenue between two parties considers business perspective where the project should meet their expectation of income and level of risk. High-speed train and tourism have the most significant income compared to the other functions; thus, 80% of the revenue will be transferred to the business entity. Although fiber optic contributes lower income in total compared to the other functions, the internal rate of return is relatively high for about 32%. The simulation shows that 50% sharing for both parties is the most suitable by considering equality among functions and minimizing monopoly of the sector. As a result, the multifunction project of high-speed train using PPP scheme produces an internal rate of return for about 16.1% with a positive net present value (NPV). The value-added project arguably produces a significant impact on increasing the infrastructure feasibility and possibility for other similar project development.

6. Institutional scheme

The institutional scheme is generated from previous financial engineering and PPP scheme of Jakarta-Surabaya HST project development. The system will administer the responsibility between government, private investors, and other parties in the project. Firstly, the companies that involved in the project consist of several backgrounds and provide distinct role and capabilities according to the five functions from value engineering process. They collaborated with the government for an agreement and supported by other entities such as contractors, consultants, donors for the whole project's lifecycle. The framework can be seen in **Figure 8**.

Investors will support the initial cost of about 142 trillion rupiahs and 9 trillion rupiahs of operation and maintenance subjected to the financial scheme. The investors might consist of domestic or foreign companies by considering their experiences, capabilities, and other related factors. On the other hand, the consultant consists of Center for Sustainable Infrastructure Development (CSID) who generates the concept and other established foreign companies to support the project's feasibility.

For the contractors, high-speed train function will be managed by an international company in collaboration with railway state-owned enterprise (SOE) who has experience in running railroad business. The agreement shall govern technology and knowledge transfer from the foreign and domestic partner during the concession period. The scheme is expected to improve capacity and capability of the SOE in developing and managing rail, signaling, and station to rolling stock of high-speed train. Other functions such as fiber optic, transit-oriented development, solar cell,

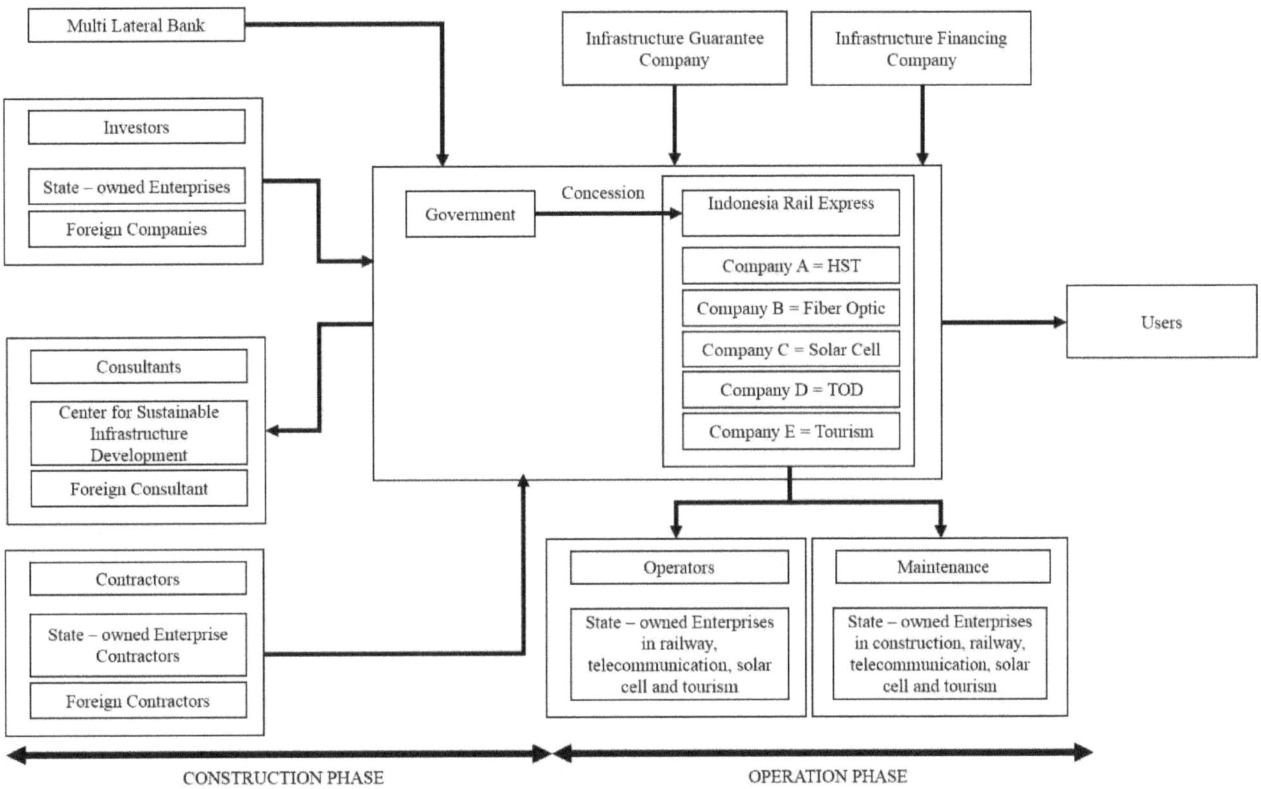

Figure 8. Institutional framework for HST project development in Indonesia.

and tourism will be selected based on their capability and innovative ways to generate revenues. The companies might be from domestic, foreign, or a partnership between both parties.

Operation and maintenance are for about 30 years. Operators are from the state-owned enterprise in collaboration with reputable international companies. Maintenance is also conducted in similar ways with the operation. In this phase, cost sharing during operation and maintenance within the joint venture will be divided equally between the government and business entity. Revenues generated from the services or product in the project will be shared—23.19% by the government and 78.81% for the business entity. The government through appointed institution collects revenues as part of the involvement in financing the initial stage and operation and maintenance stage. The scheme generates optimal revenues for business and government acquires income, which can be used to build other infrastructure or support local transportation through subsidy.

7. Conclusion and progress of the project

Value engineering implementation in the HST project development has transformed a single-function project from transporting people into a multifunction project. Additional functions that have been identified are transit-oriented development (TOD), fiber optic, tourism, and solar cell.

Added value from the innovative function enables the project to reach an expected value of feasibility through significant internal rate of return (IRR) and positive net present value

(NPV). Before value engineering method, the single-function project only reaches 5.20% of IRR. Meanwhile, the additional function has been successfully improving the IRR value into 12.30%. Furthermore, public-private partnership scheme elaborates the division of responsibility between government and business entity about financing the project. The result increases the IRR value to 16.1% Thus, a collaboration between value engineering and the public-private partnership is a package to improve project feasibility particularly railway development to reach its expected outcome from stakeholders' perspective.

Currently, high-speed train project in Indonesia is in the construction phase. However, the progress remains low below 5% in the past 2 years due to land acquisition, funding, and institutional scheme. The project needs 700 ha for land acquisition and currently achieved 54.5%. Land owned by the state-owned enterprise and government institution is relatively more accessible for acquisition due to robust communication and coordination among them. However, the private property is required further negotiation. The project has been receiving a loan from Chinese bank—about 13 trillion rupiahs or equal to US$ 1 billion. The funding attempt to support the consortium consists of Indonesia's and Chinese state-owned enterprises in accelerating project realization.

Acknowledgements

This research is supported by research grants from the Ministry of Research, Technology and Higher Education and Universitas Indonesia.

Author details

Mohammed Ali Berawi

Address all correspondence to: maberawi@eng.ui.ac.id

Faculty of Engineering, Universitas Indonesia, Indonesia

References

[1] Seneviratne MD, Sun MY. Infrastructure and Income Distribution in ASEAN-5: What are the Links? International Monetary Fund; 2013

[2] Das SB, Menon J, Severino RC, Shrestha OL. The ASEAN Economic Community: A Work in Progress. 14th ed. Singapore: Institute of Southeast Asian Studies; 2013

[3] Schwab K. The Global Competitiveness Report. World Economic Forum; 2012

[4] Dikun, S. The Interface Report: Substances to Support the National Railway Master Plan. 2010

[5] Berawi MA, Berawi A, Prajitno I, Nahry N, Miraj P, Abdurachman Y, Tobing E, Ivan A. Developing conceptual design of high speed railways using value engineering method: Creating optimum project benefits. International Journal of Technology. 2015;**6**(4):670-679. DOI: 10.14716/ijtech.v6i4.1743

[6] Berawi MA, Miraj P, Berawi ARB, Sumabrata J. Track access charge for Indonesian railways using full cost method: Improving industry competitiveness. MATEC Web of Conferences. 2017;**101**:05002

[7] Levinson DM. Accessibility impacts of high-speed rail. Transport Geography. 2012; **22**:288-291

[8] Berawi MA. Distinguishing Concept Types in Function Models During the Act of Innovation [Dissertation]. Oxford: Oxford Brookes University; 2006

[9] Miles LD. Techniques of Value Analysis and Engineering. Miles Value Foundation; 2015. https://books.google.co.id/books?hl=en&lr=&id=yCf0CQAAQBAJ&oi=fnd&pg=PT16&dq=Techniques+of+Value+Analysis+and+Engineering.+Miles+Value+Foundation%3B+201&ots=vVpyiHULCv&sig=mUviX-dlZavMvkRj4qlXK8HKN1M&redir_esc=y#v=onepage&q=Techniques%20of%20Value%20Analysis%20and%20Engineering.%20Miles%20Value%20Foundation%3B%20201&f=false [Online]

[10] Woodhead RM, Berawi MA. An alternative theory of idea generation. International Journal of Management Practice. 2008;**3**(1):1-9

[11] Berawi MA, Woodhead RM. Application of knowledge management in production management. Human Factors and Ergonomics in Manufacturing & Service Industries. 2005;**15**(3):249-257

[12] Chen WT, Chang PY, Huang YH. Assessing the overall performance of value engineering workshops for construction projects. International Journal of Project Management. 2010;**28**(5):514-527

[13] Lindahl M. Track geometry for high-speed railways. TRITA-FKT Report. 2001;34

[14] Daddow T, Skitmore M. Value management in practice: An interview survey. The Australian Journal of Construction Economics and Building. 2005;**4**(2):11-18

[15] Zhang X, Mao X, AbouRizk SM. Developing a knowledge management system for improved value engineering practices in the construction industry. Automation in Construction. 2009;**18**(6):777-789

[16] Ibusuki U, Kaminski PC. Product development process with focus on value engineering and target-costing: A case study in an automotive company. International Journal of Production Economics. 2007;**105**(2):459-474

[17] Berawi MA, Woodhead RM. Stimulating innovation using function models: Adding product value. Value World. 2008;**31**(2):4-7

[18] Tohidi H. Review the benefits of using value engineering in information technology project management. Procedia Computer Science. 2011;**3**:917-924

[19] Bodewes WE. Formalization and innovation revisited. European Journal of Innovation Management. 2002;**5**(4):214-223

[20] Cumming BS. Innovation overview and future challenges. European Journal of Innovation Management. 1998;**1**(1):21-29

5

Railway's Stability Observation by Satellite Radar Images

Davod Poreh, Antonio Iodice, Daniele Riccio and Giuseppe Ruello

Abstract

Remote sensing has many vital civilian applications. Space-borne Interferometric Synthetic Aperture Radar has been used to measure the Earth's surface deformation widely. In particular, Persistent Scatterer Interferometry (PSI) is designed to estimate the temporal characteristics of the Earth's deformation rates from multiple InSAR images acquired over time. This chapter reviews the space-borne Differential Interferometric Synthetic Aperture Radar techniques that have shown their capabilities in monitoring of railways displacements. After description of the current state of the art and potentials of the available radar remote sensing techniques, one case study is examined, pertaining to a railway bridge in the Campania region, Italy.

Keywords: railways deformation, synthetic aperture radar (SAR), interferometry, persistent scatterer interferometry (PSI), Cosmo-SkyMed (CSK)

1. Introduction

Radar waves are not influenced significantly by clouds and/or atmosphere, except for very heavy rain and tornados, etc.; therefore, it is a unique engineering tool for monitoring the Earth's surface. One of the brilliant inventions in the field of satellites radar remote sensing is known as persistent scatterer interferometry (PSI), which employs synthetic aperture radar (SAR) image time series. PSI is an advanced, more recent implementation of the differential interferometric synthetic aperture radar (DInSAR) technique. PSI is a powerful approach that has been employed for more than 15 years for studying/monitoring of deformation rates of man-made structures. It uses satellite interferometric synthetic aperture radar (InSAR) imageries as input. The PSI methodology is particularly suitable to urban/man-made regions, rather than to rural areas where the coherency of the radar signals is decreased dramatically. In the traditional DInSAR data analysis methodologies, coherency, geometrical decorrelation, and phase delay due to atmospheric effects on electromagnetic (EM) waves, are three major limitations of the application. However, for some specific conditions like urban areas, highways, or railways, temporal decorrelation decreases dramatically (i.e., coherency is higher), and features remain coherent in the produced interferograms for a long time [1–5]. To overcome the coherency problems of backscatterers (changes of the backscatters during the time) in repeat pass SAR interferometry, PSI was developed.

Historically speaking, different PSI techniques have been proposed in last decades. The first PSI technique was developed by researchers of the Politecnico di Milano (POLIMI) in Italy [1, 2]. Soon after, some other similar methodologies have been rapidly developed. The other similar well-known time series radar interferometric approach is named small baseline subset (SBAS) [6–9]. In PSI analysis, all acquisitions are employed, whereas in SBAS some of them are not, because their spatial baseline is too high. SBAS methodologies are more sensitive to geometric and temporal decorrelation compared to PSI analysis [6, 7].

In SBAS, much more interferograms are created than in a single-master approach (like PSI). The unwrapping procedure for SBAS and PSI is also different. In SBAS, at least in its original implementation, the interferograms are unwrapped first spatially and then temporally, while it is the opposite in the PSI analysis. One of the major disadvantages of SBAS is that in this approach disconnected clusters of interferograms might be obtained in the temporal and perpendicular baseline graphs. However, SBAS allows to measure displacements not only on highly stable point-like scatterers, but also on distributed scatterers (DS), i.e., areas with intermediate coherence. Therefore, several researches have been reported aiming to develop techniques that are able to combine advantages of both PSI and SBAS. For instance, minimization of the baselines and use of all radar images also in SBAS methodology were proposed in the literature, e.g., [10]. Another similar technique for monitoring the Earth's surface change was reported in [11]. A geophysical approach in [12] and a stepwise linear deformation with least square adjustment in [13] has been reported. Interferometric point target analysis (IPTA) and stable point networks are reported in [14, 15], respectively. In [3], multiple image pixels are used within a certain radius to estimate spatially correlated parameters (e.g., deformation rates and atmospheric signal delay). In this methodology, PSI and small baseline analysis have been combined heuristically. The SqueeSAR (Squeeze more information from SAR images) algorithm, capable of simultaneous analysis of PSI (i.e., PS) and DS, was reported in [16]. In SqueeSAR algorithm, combination of PS and DS helps to work in rural areas, where the coherency is lower. In [17], a similar algorithm, with contribution of polarimetric-based radar data, was heuristically proposed.

In PSI implementations, a large stack of radar images are considered for the estimation of historical changes of the Earth's surfaces, with proper modeling techniques. The output of PSI algorithms are deformation time series of the scatterers, and the elevation of those scatterers. PSI technique exploits the fact that some radar's pixels remain coherent during the time. With this method and by using a large stack of SAR images (usually more than 20 SAR images), atmospheric errors (i.e., the Atmospheric Phase Screen, APS) can be estimated with sufficient accuracy, and the proper phase correction can be implemented to remove them. In the standard PSI methodologies, a single master image with specific criteria is selected (from N given images), and $(N - 1)$ differential interferograms w.r.t. the master image are generated. Then, with different approaches, permanent scatterers candidates (PSCs) are selected. By refinements of the selected PSCs, and by using permanent scatterers potentials (PSPs) [18], final permanent scatterers (PSs) can be generated. For each radar permanent/persistent Scatterer point, time series of the historical records of the Earth surface's height changes, and the height of each PS with respect to a reference point, are measurable. This methodology shows promising results in urban areas, where it is able to achieve an average of 100 PSs/km^2 (points densities) with low resolution sensors like ERS1/2 and ENVISAT-ASAR, and an average of a couple of thousands PSs/km^{2}with high resolution sensors like TerraSAR-X (TSX) and Cosmo-SkyMed. On the contrary, the rural/vegetated areas might not be explored properly with PSI methodology. The main reason is the absence of proper stable scatterers in such areas. Another disadvantage of the PSI is the need for a minimum amount of images for making appropriate phase unwrapping steps, which could severely influence the degree of correctness of the selected PSC. The other limitation of InSAR time series methodologies, is that PSI (and SBAS too) is a relative technique, i.e., all of the calculated time series for PS points are measured w.r.t. a reference point, which is assumed to be without any kind of movements. Nonetheless, many promising methodologies, like continuous GPS or leveling, could resolve this problem properly [18]. Another limitation is mainly due to the observation geometry of the satellite systems. PSI deformation rates are only measured along the satellite line of sight (LOS) direction; therefore, the obtained value of the deformation is actually just the projection of the deformation vector onto the SAR look direction. One example of PSI analysis is given in **Figure 1**. Fifty-two ENVISAT-ASAR imageries in descending mode over the Mexico City region have been analyzed, and the deformation rates over this area have been presented. Nine GPS stations that have been installed in this area to help displacement retrieval, are also depicted in this picture.

2. Engineering application review

In this section, we review some of the most important engineering applications of radar monitoring, with main focus on transportation infrastructure control. Several projects dealing with SAR and InSAR applications have been and are currently being carried out by both academic and research groups and by specialized companies: they are described in the scientific literature, as listed below, and on companies' websites (see, e.g., www.altamira-information.com, www.gamma-rs.ch, and www.npagroup.com).

One of the most important and elegant applications of the radar monitoring is about urban areas. With radar images, huge areas can be covered/monitored with high-resolution images,

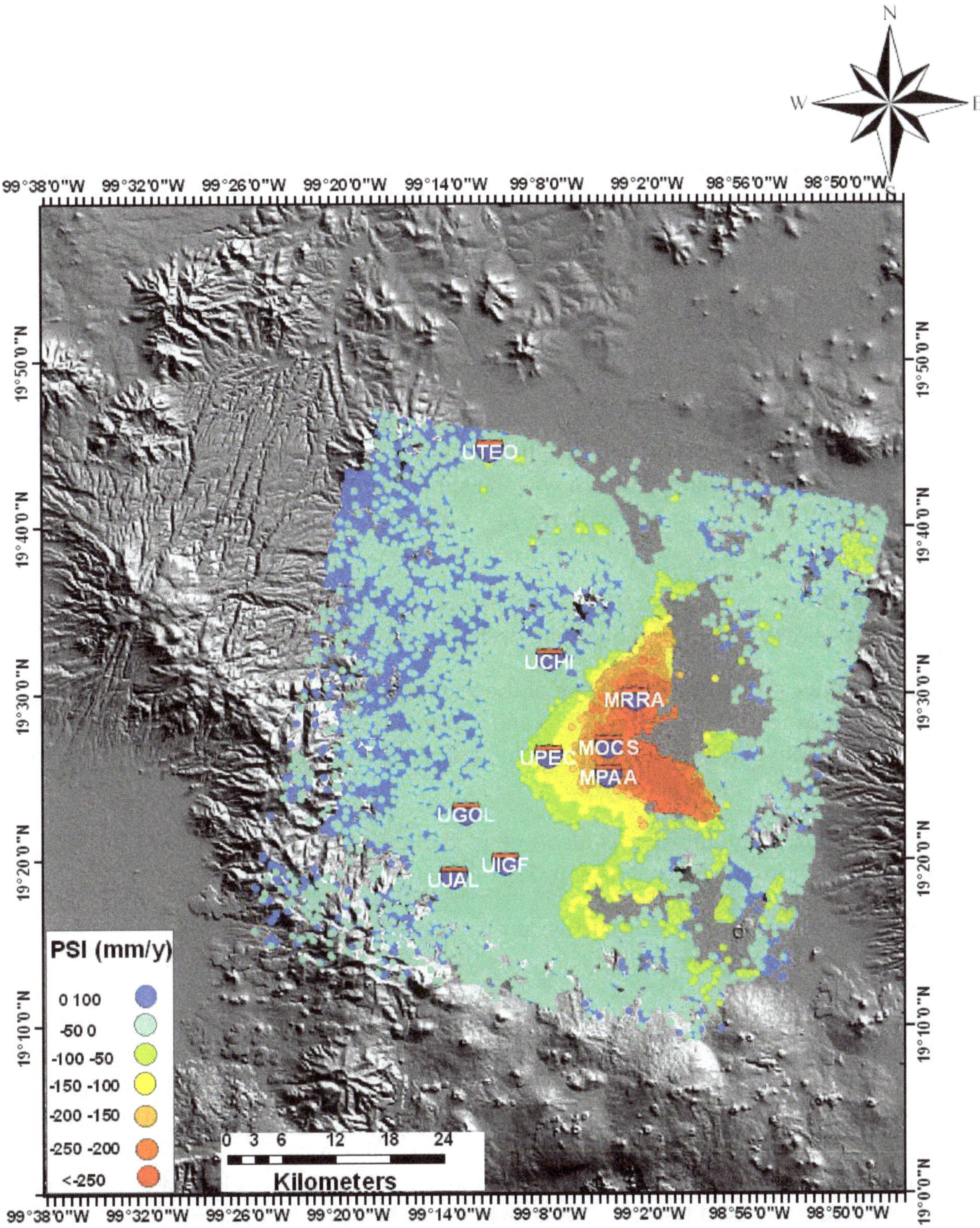

Figure 1. Example of PSI deformation velocity map over the Mexico City area. Fifty two ENVSAT-ASAR imageries and nine (permanent) GPS stations that acquired between 2002- June, 2010 and 1998-2012, respectively are analyzed and the results are depicted in this figure. The eastern part of Mexico City metropolitan area, has been subsiding at fast and constant rates for decades. Maximum continued subsidence rate of 352 mm/year in LOS direction is happening in the east and central part of Mexico City metropolitan area (Radar images are from ESA, and background SRTM (Shuttle Radar Topography Mission) images are from NASA).

and with acceptable re-visiting schedules. Historical measurements of the urban area's (at least from 1992) deformations and the cost-effectiveness of the monitoring are two important advantages of satellite radar monitoring in the human-made objects.

Radar remote sensing of the Earth's surface subsidence and uplifting is currently used in several cites all around the world. For instance, in the PanGeo project (http://www.pangeoproject.eu/eng/project_overview), more than 52 cities of the Europe which host around 13% of the entire populations, were under this kind of surveillance continually. This project gives online information to users for possible geo-hazard evaluations in their cities. However, the radar images that this project used are more or less low-resolution imageries, and no detail information could be retrieved. For instance, these results might not be very useful for detailed railways and bridges benefits.

For a detailed study of single target with radar images, one can refer to [19], where subsidence of the New Orleans has been studied. Another brilliant example is given by Zerbini et al. [20], where the Bologna region and Po Plain was the subject of the study. An example of subsidence monitoring due to water extraction is given by Osmanoglu et al. [21] for Mexico City metropolitan area.

Telerilevamento Europa (www.treuropa.com) runs a big project for subsidence monitoring due to oil and gas extractions that might be useful for oil and gas companies.

Faults movements monitoring is also one of the prominent usage of the radar remote sensing techniques. One example is given by Lyons and Sandwell [22]. With satellite radar monitoring of the faults, the rate of the creeping along the fault lines could be measured, and this might be translated to fault's future activities.

Landslide geo-hazard monitoring is also one of the most considered subjects in the radar remote sensing, for instance, see Refs. [23, 24].

In the Netherlands, radar remote sensing of dikes and dams for water defense systems is a common task, for instance, see [25].

Let us now consider the available applications on monitoring of transportation infrastructures (particularly, railways). In [26], a bridge displacement with use of very high (spatial) resolution X-band TerraSAR-X imageries is examined. The performed analysis shows that horizontal movement of the bridge is due to thermal expansion of the bridge and the vertical deformation is due to the settlements (during the time) of the bridge itself. The authors of that study proposed a combined model for thermal and (InSAR) phase residual rates estimations. They claimed that at least 1.5-year-cycle images are needed to reach a reliable estimation of deformations with InSAR techniques.

Finally, it is worth mentioning that [27] used RADARSAT-2 images for monitoring of the railways instabilities in the entire Netherlands. They use 73 images acquired from 2010 to 2015 to give a clear perspective from deformation scheme of the railways, and with some statistical approaches they evaluate the quality of the PSI analysis.

3. A case study of monitoring of railways stability in Italy (Campania region)

In this section, the results of the DInSAR PS technique to the remote monitoring of railways in the Campania region in Italy (**Figure 2**) are reported. In particular, we are interested in monitoring a bridge over the Volturno river, at Triflisco. As widely reported in literature [28–32], Campania region is very unstable in terms of the Earth's surface deformations. Intense urbanization, active volcanoes, complicated fault systems, landslides, subsidence, and hydrological instability (flooding) are the characteristics of this region [28–32]. 246 out of 652 sinkholes (38%) of the entire Italy are located in Campania region itself [31]. Volcanism is very developed in this area and is observed at Roccamonfina, Ischia, Vesuvius, and Phlegraean Fields regions. Historical eruptions happened at Vesuvius (several, the last one in 1944 A.D.), Ischia (1302 A.D.), and Phlegraean Fields (1538 A.D.). Therefore, this area is under geo-based investigation periodically. In addition, in the Campania region, the railways and bridges are pretty old, and are prone to sudden or slow deformation threats. For instance, the bridge over the Volturno river and the railways considered in this study were made in 1953.

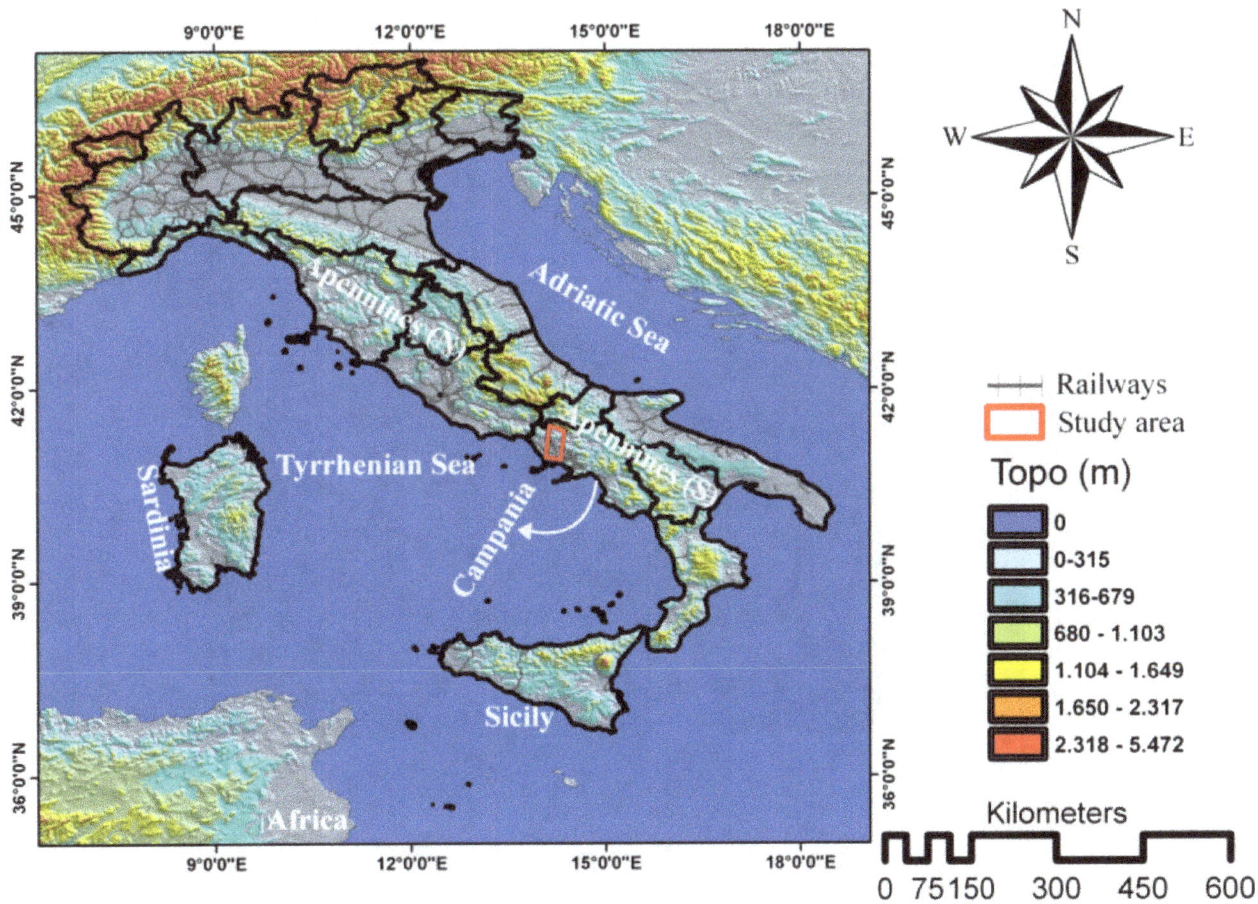

Figure 2. Study area and railways networks are depicted in this figure.

We were interested in the area at north of Napoli, including the Volturno river and a local railway, see **Figure 3**, with a bridge at Triflisco on which our interest is mainly focused. This bridge over the Volturno river and the entire railway considered in this study were made in 1953. In order to increase the stability of this bridge, the local railway company (EAV) made some rock bolts installations and cement injections to make the three pillars of the bridge more stable. Despite these efforts, EAV still was interested to evaluate the probable geophysical change of the railways (deformation rates) with other independent methods such as InSAR and PSI.

The study area has already gone under PSI analysis with low-resolution images like ERS, and ENVISAT-ASAR from PlaneTek Company before [33]. However, with these sensors, the number of PSs in the study area is too small. For instance, on the bridge over the Volturno river considered in this study, with ERS data sets from 1992 to 2000, in the ascending mode, only two, and, with the descending mode, only one PS have been selected. With ENVISAT-ASAR sensor in temporal baseline of 2003–2010, in the ascending mode six, and in the descending mode only one PS have been selected. Therefore, the need for using high-resolution imageries for a better understanding of the deformation phenomena on the bridge is obvious. In [33], first results on Cosmo-SkyMed data are reported, obtained by using the SPINUA (Stable Point Interferometry over Un-urbanised Areas) algorithm. We more recently performed a new analysis [34] by using a different algorithm [18]. Results of this analysis are recalled in the following.

Twenty-five InSAR images of Cosmo-SkyMed sensors at descending mode of HIMAGE/ Stripmap were used for the considered study area, (it is depicted in **Figure 2** as a red rectangle). Images are acquired in HH polarization, right looking, X-band (EM wavelength: 3.1228 cm), with mean incident angle of 26.60° (incidence angle at the center of the transmitted beam).

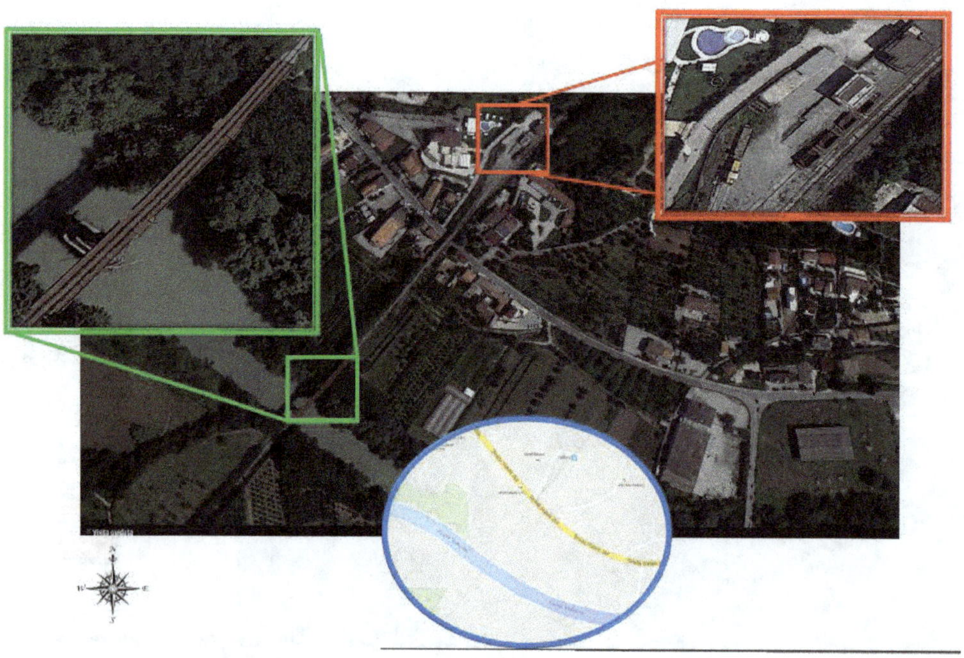

Figure 3. Study area, railways, targeted bridge, and the Volturno river are depicted in this figure. Down in the middle a global view of the study area is depicted (from Google Maps). On top left, the targeted bridge of the Volturno river, and on the top right, the train station are enlarged.

Data cover a temporal baseline between February 24, 2011 and March 23, 2015. We examine this stack of images to identify the number of scatterers on the ground that consistently/permanently show stable reflections back to the satellite on all images in the temporal baseline. With PSI analysis, historical motion of the permanent scatterers on the ground was determined. Image of June 5, 2013 has been selected as master image, and radar images were cropped in an area as big as 7.5 × 7.5 km^2, centered at the bridge on the Volturno river (**Figure 3**). Then, 24 differential interferograms have been generated w.r.t. the master image. With Cosmo-SkyMed data sets and for the selected study area of 7.5 × 7.5 km^2, more than 190,000 PSs including some on the railways, and some on the bridge of the Volturno river have been selected. The average velocity and ensemble coherence are as big as −1.8 mm/year (for whole area) and 73%, respectively, and the density of selected PSs is equal to 3378 PSs/km^2 for the entire area.

The majority of the PSs are from man-made structures such as houses, highways, railways, etc. **Figure 4** shows the mean velocity (deformation rate) of the Earth's surface in the study area. As it is clear from this figure, man-made structures such as highways, railways, and cities are designated as potential permanent radar wave reflectors (i.e., PSs).

Figure 5 shows the selected scatterers (PSs) on the railways and the bridge itself. It turns out that they are 1385 and, as it is obvious from this figure, most of them are stable. Minimum and maximum displacement rates are −9.58 and 9.97 mm/year, respectively, but these high values are only obtained in some isolated points (dots in **Figure 5**), surrounded by points for which the displacement rates are much lower, so that they are probably due to phase noise.

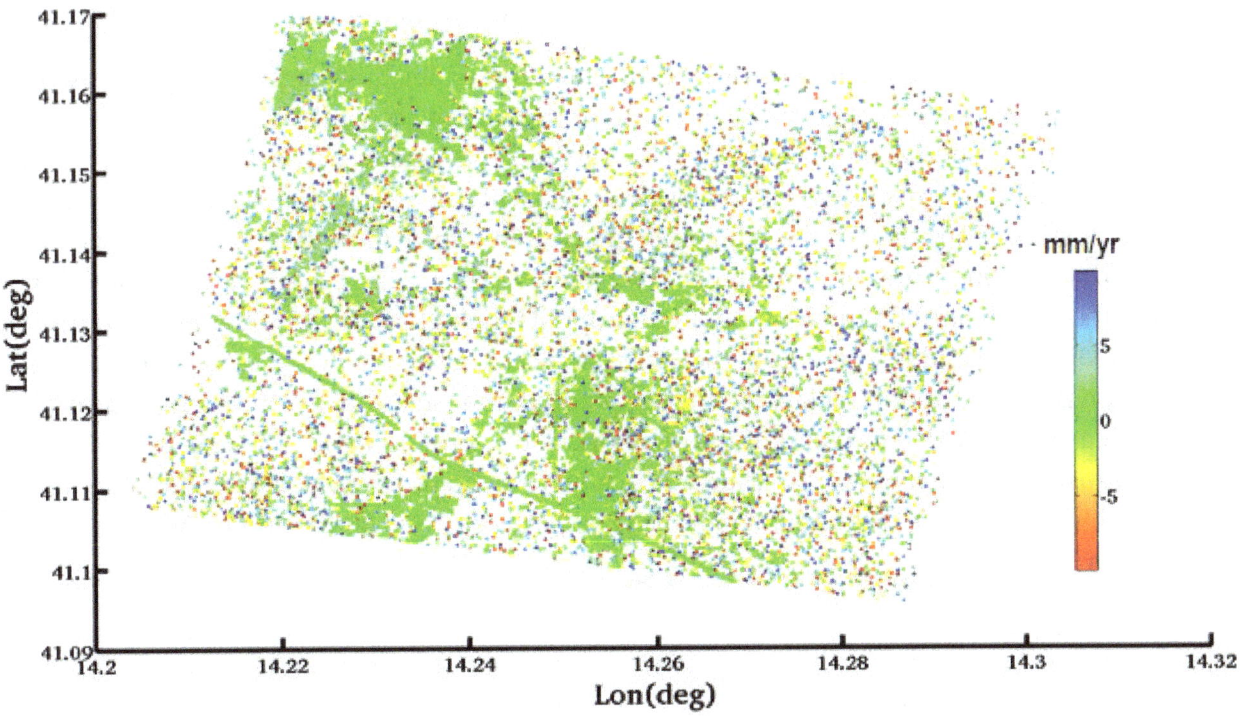

Figure 4. Mean surface displacement velocity of the study area based on 25 InSAR images from 2011 02 24 until 2015 03 23, estimated with PSI methodology. Gray pixels are PSs, white pixels are not PSs and no velocity estimation is obtained for them (for more details see [34]).

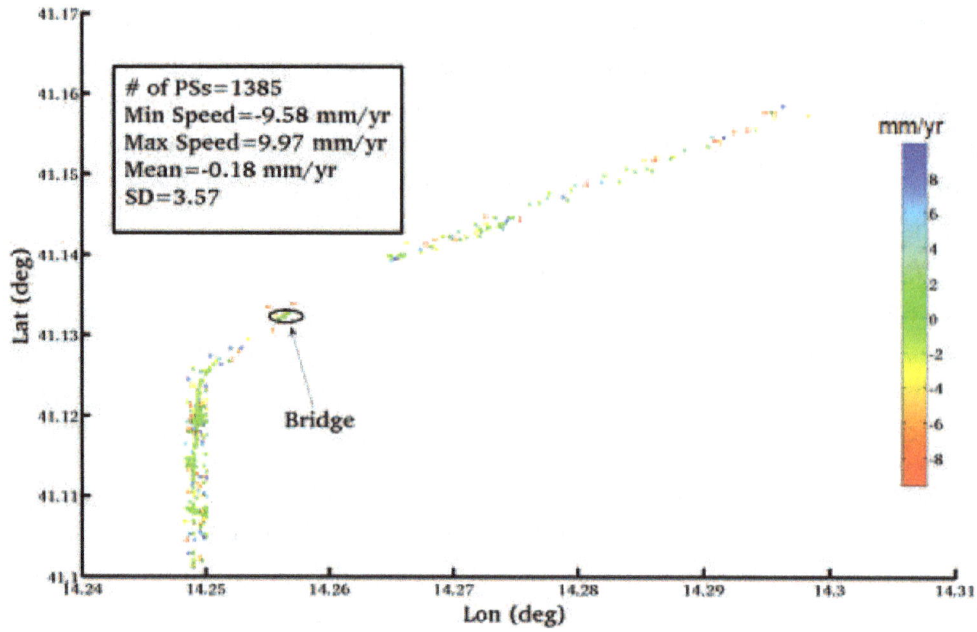

Figure 5. Mean displacement rates of PSs on the railways and statistics of these PSs (for more details see [34]).

The velocity averaged over all the 1385 PSs is 1.8 mm/year, and standard deviation (SD) is 3.57 mm/year. The PSs on the railways and nearby structures are selected in the GIS environment manually. Thirty of these PSs are located on the bridge, and they are highlighted by an ellipse in **Figure 5**. For these 30 PSs, minimum and maximum velocities of −0.9 and 0.05 mm/year have been observed, respectively, with an average of −0.3 mm/year and SD = 0.3 mm/year.

For each PS, not only the displacement rate, but also the entire time series of displacements is obtained. This allows for a deeper analysis of the bridge displacements' behavior and identifying probable acceleration along the temporal baselines. In particular, we have compared the time variations of the bridge displacement (or deformation) with the time variations of temperature in the same considered area and in the same time interval. In **Figure 6**, the points connected by lines show the deformation time series averaged over the 30 PSs' on the bridge, and the dots points represent the temperature in the Neapolitan metropolitan area [35]. As it is obvious from this figure, the deformation and temperature time series are very similar, demonstrating that most of the deformation is cyclical and it is related to the temperature seasonal changes in winter and summer time. Decreasing of the detected amplitude of deformation yearly cycle in 2013 and 2014 (**Figure 6**) is actually probably due to undersampling: in fact, it is related to the smaller number of images in that period, with no image acquired in summer 2013 and only one (at the end of August) in summer 2014.

In conclusion, comparison of average PSs time series on the bridge with thermal data shows that most of the line of sight (LOS) changes are due to the periodical variations of the temperature (i.e., winter and summer), with cyclical, seasonal deformations superimposed to a small rate of deformation of −0.30 mm/year.

Accordingly, the use of higher resolution imageries like Cosmo-SkyMed (CSK) and TSX to have better and smooth time series has been demonstrated by using CSK data. However,

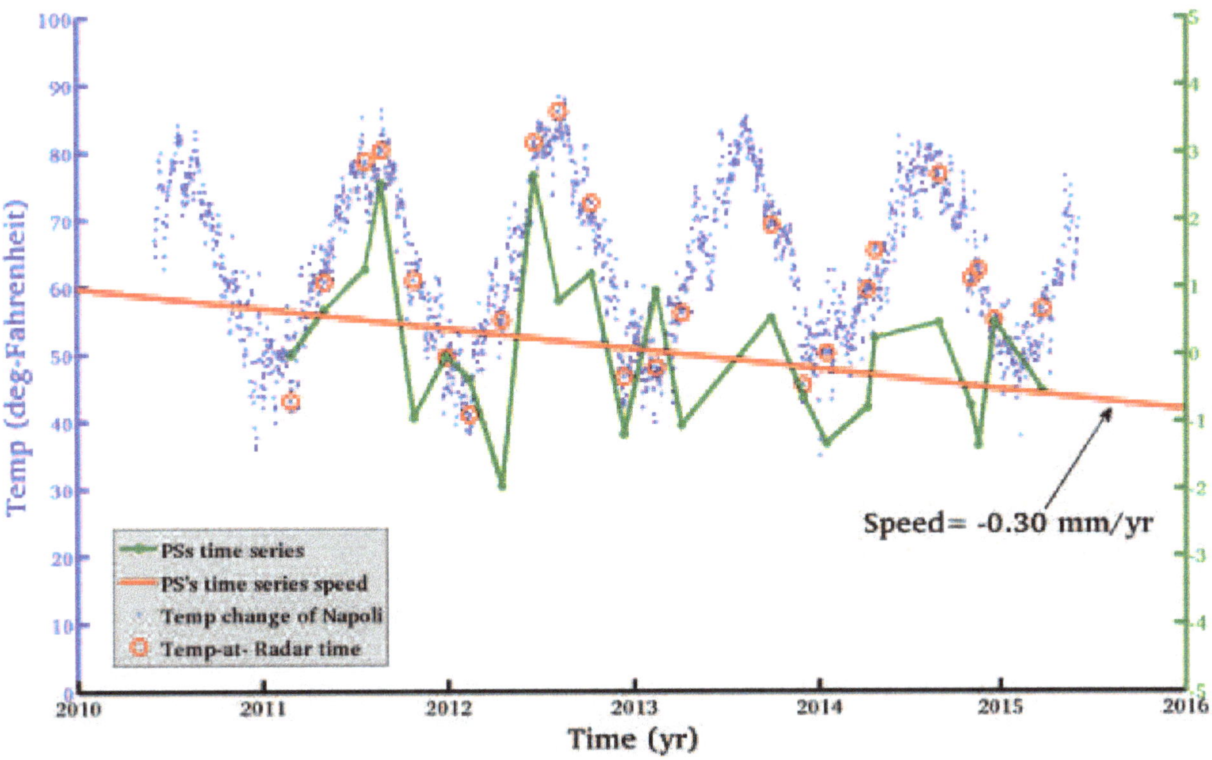

Figure 6. Bridge deformation as a function of time (points connected by lines) compared with temperature of the Neapolitan metropolitan area as a function of time (dots). Bridge deformation is obtained by averaging the values of all the 30 PSs on the bridge, and temperature data are taken from NOAA (for more details see [34]).

the combination of descending mode with ascending mode imageries to achieve the deformation rates in both vertical and horizontal directions (i.e., not only along the line of sight), continuous GPS deformation monitoring, corner reflector establishment, and leveling data might improve understanding of this study area. Comparison of SBAS methodology with the employed PSI technique also would be helpful and is the subject of our group's assignment.

4. Conclusion

In this chapter, the monitoring of railways based on the radar remote sensing and mainly PSI technique has been discussed. The key characteristics of InSAR-based technique have been described and benefits/disadvantages of the techniques have been highlighted. The main products of radar remote sensing surveillance have been briefly described and some of the most important projects have been concisely reviewed, providing a comprehensive list of references.

An important point that we stress here is that the availability of high-resolution SAR data is a key need for interferometric approaches if we want to monitor deformation rates of infrastructures like railways. In addition, it should be pointed out that, due to limited (re)visiting and SAR viewing geometries, PSI is not always capable of providing a real-time warning of possible critical deformations that might be occurred on railways, so that for this latter aim PSI should be employed in conjunction with other monitoring methods.

Author details

Davod Poreh*, Antonio Iodice, Daniele Riccio and Giuseppe Ruello

*Address all correspondence to: davod.poreh@unina.it

Università degli Studi di Napoli "Federico II", Dipartimento di Ingegneria Elettrica e delle Tecnologie dell'Informazione, Napoli, Italy

References

[1] Ferretti A, Prati C, Rocca F. Permanent scatterers in SAR interferometry. IEEE Transactions on Geoscience and Remote Sensing. 2001;**39**(1):8-20

[2] Ferretti A, Prati C, Rocca F. Nonlinear subsidence rate estimation using permanent scatterers in differential SAR interferometry. IEEE Transactions on Geoscience and Remote Sensing. 2000;**38**:2202-2212

[3] Ferretti A, Prati C, Rocca F. Non-uniform motion monitoring using the permanent scatterers technique. In: Second International Workshop on ERS SAR Interferometry, 'FRINGE99 conference', Liege, Belgium, 10-12 November 1999, ESA. 1999. pp. 1-6

[4] Ferretti A, Prati C, Rocca F. Multibaseline InSAR DEM reconstruction: The wavelet approach. IEEE Transactions on Geoscience and Remote Sensing. 1999;**37**(2):705-715

[5] Usai S, Hanssen R. Long time scale INSAR by means of high coherence features. In: Third ERS Symposium – Space at the Service of our Environment, Florence, Italy, March 17-21, 1997. 1997. pp. 225-228

[6] Berardino P, Fornaro G, Lanari R, Sansosti E. A new algorithm for surface deformation monitoring based on small baseline differential SAR interferograms. IEEE Transactions on Geoscience and Remote Sensing. 2002;**40**:2375-2383

[7] Lanari R, Mora O, Manunta M, Mallorqui JJ, Berardino P, Sansosti E. A small-baseline approach for investigating deformations on full-resolution differential SAR interferograms. IEEE Transactions on Geoscience and Remote Sensing. 2004;**42**:1377-1386

[8] Pepe A, Manunta M, Mazzarella G, Lanari R. A space-time minimum cost flow phase unwrapping algorithm for the generation of persistent scatterers deformation time-series. In: Proceedings of the IEEE International Geoscience and Remote Sensing Symposium (IGARSS 2007), Barcelona, Spain, July 23-28, 2007. 2007

[9] Pepe A, Sansosti E, Berardino P, Lanari R. On the generation of ERS/ENVISAT D-InSAR time-series via the SBAS technique. IEEE Geoscience and Remote Sensing Letters. 2005;**2**:265-269

[10] Pepe A, Berardino P, Bonano M, Euillades LD, Lanari R, Sansosti E. SBAS-based satellite

orbit correction for the generation of DInSAR time-series: Application to RADARSAT-1 data. IEEE Transactions on Geoscience and Remote Sensing. 2011;**49**:5150-5165

[11] Mora O, Mallorqui JJ, Broquetas A. Linear and nonlinear terrain deformation maps from a reduced set of interferometric SAR images. IEEE Transactions on Geoscience and Remote Sensing. 2003;**41**:2243-2253

[12] Hooper A, Zebker H, Segall P, Kampes B. A new method for measuring deformation on volcanoes and other natural terrains using InSAR persistent scatterers. Geophysical Research Letters. 2004;**31**:1-5. DOI: 10.1029/2004GL021737

[13] Crosetto M, Crippa B, Biescas E. Early detection and in-depth analysis of deformation phenomena by radar interferometry. Engineering Geology. 2005;**79**:81-91

[14] Werner C, Wegmüller U, Strozzi T, Wiesmann A. Interferometric point target analysis for deformation mapping. In: Proceedings of the 2003 Geoscience and Remote Sensing Symposium (IGARSS 2003), Toulouse, France, July 21-25, 2003. 2003

[15] Crosetto M, Biescas E, Duro J, Closa J, Arnaud A. Generation of advanced ERS and ENVISAT interferometric SAR products using the stable point network technique. Photogrammetric Engineering and Remote Sensing. 2008;**74**:443-450

[16] Ferretti A, Fumagalli A, Novali F, Prati C, Rocca F, Rucci A. A new algorithm for processing interferometric data-stacks: SqueeSAR. IEEE Transactions on Geoscience and Remote Sensing. 2011;**49**:3460-3470

[17] Navarro-Sanchez VD, Lopez-Sanchez JM. Spatial adaptive speckle filtering driven by temporal polarimetric statistics and its application to PSI. IEEE Transactions on Geoscience and Remote Sensing. 2013;**99**:1-10

[18] Kampes B. Displacement parameter estimation using permanent scatterer interferometry. Ph.D. thesis, Technische Universiteit Delft. 2005

[19] Dixon TH, Amelung F, Ferretti A, Novali F, Rocca F, Dokka R, et al. Space geodesy: Subsidence and flooding in New Orleans. Nature. 2006;**441**(7093):587-588. DOI: 10.1038/441587a

[20] Zerbini S, Richter B, Rocca F, van Dam T, Matonti F. A combination of space and terrestrial geodetic techniques to monitor land subsidence: Case study, the Southeastern Po Plain, Italy. Journal of Geophysical Research. 2007;**112**(B05401):1-12

[21] Osmanoglu B, Dixon TH, Wdowinski S, Cabral-Cano E, Jiang Y. Mexico City subsidence observed with persistent scatterer InSAR. International Journal of Applied Earth Observation and Geoinformation. 2011;**13**(1):1-12

[22] Lyons S, Sandwell D. Fault creep along the southern San Andreas from interferometric synthetic aperture radar, permanent scatterers, and stacking. Journal of Geophysical Research. 2003;**108**(B1). DOI: 10.1029/2002JB001831

[23] Colesanti C, Wasowski J. Investigating landslides with space-borne Synthetic Aperture

Radar (SAR) Interferometry. Engineering Geology 2006;**88**:173-199

[24] Kimura, H, Yamaguchi Y. Detection of landslide areas using satellite radar interferometry. Photogrammetric Engineering & Remote Sensing. 2000;**66**(3):337-344

[25] Hanssen RF, van Leijen FJ. Monitoring water defense structures using radar interferometry. In: Radar Conference, May 26-30, 2008. Radar '08. Rome, Italy: IEEE; 2008

[26] Lazecky M, Hlavacova I, Bakon M, Sousa JJ, Perissin D, Patricio G. Bridge displacements monitoring using space-borne X-band SAR interferometry. IEEE Journal of Selected Topics in Applied Earth Observations and Remote Sensing. 2017;**10**(1):205-210

[27] Hanssen R, Chang L Dollevoet RPBJ. Automatic Railway Instability Detection Using Satellite SAR Interferometry, MUAS-2015, November 4, ESA Frascati.

[28] Blonda P, Parise M, Reichenbach P, Guzzetti F. Rock-fall hazard assessment along a road in the Sorrento Peninsula, Campania, southern Italy. Natural Hazards. 2012;**61**(1):187-201

[29] Calcaterra M, Parise M, Blonda P. Combining historical and geological data for the assessment of the landslide hazard: A case study from Campania, Italy. Natural Hazards and Earth System Sciences. 2003;**3**(1/2):3-16

[30] Del Prete S, Di Crescenzo S, Santangelo N, Santo A. Collapse sinkholes in Campania (Southern Italy): Predisposing factors, genetic hypothesis and susceptibility. Geomorphology. 2010;**54**:259-284

[31] Parise M, Vennari CA Chronological catalogue of sinkhole in ITALY: The first step toward a real evaluation of the sinkhole hazard. In: 13th Sinkhole Conference NCKRI SYMPOSIUM 2013. 2013

[32] Vilardo G, Ventura G, Terranova C, Matano F, Nardo S. Ground deformation due to tectonic, hydrothermal, gravity, hydrogeological, and anthropic processes in the Campania Region (Southern Italy) from permanent scatterers synthetic aperture radar interferometry. Remote Sensing and Environments. 2009;**113**(1):197-212

[33] Nutricato R, Nitti DO. Processamento di scene Cosmo-SkyMed con tecniche di interferometria differenziale PS. Internal report in Italian. 2015

[34] Poreh D, Iodice A, Riccio D, Ruello G. Railways' stability observed in Campania (Italy) by InSAR data. European Journal of Remote Sensing. 2016;**49**:417-431

[35] http://www7.ncdc.noaa.gov/CDO/cdoselect.cmd?datasetabbv=GSOD&countryabbv=&georegionabbv=

6

Concurrent Engineering Implementation in Design-Build Railway Projects

Ade Ogunsola

Abstract

Design-build as a procurement method is increasingly being used in the design and construction of greenfield rail networks, and that is despite the complexities that characterise rail networks—rail infrastructure projects involves significantly more complex systems such as safety, telecommunications, signalling and electrification. One of the key drivers for this choice of procurement method for the delivery of rail networks is that the design-build contractor commits to an aggressive schedule and implements strategies to enable the works to be completed to time and cost. One of such strategies is the application of concurrent engineering principles to the design and construction works. This Chapter gives an overview of concurrent engineering as applicable to design-build rail projects, focusing mainly on the design as an activity. It identifies factors that impact the application of concurrent engineering as well as mitigations that can be applied for the successful application of concurrent engineering principles in design-build rail projects.

Keywords: design and build, rail, overlapping, concurrent engineering, sequential logic

1. Introduction

Rail transportation is perhaps the most dependable form of transportation and thus it is not surprising that over the last decade there has been a steady growth in the number of green field rail projects. This is particularly true in Asia, Africa and the Middle East where new rail networks have been commissioned and new rail networks are being designed and constructed. Railways are complex distributed systems with capital expenditure ranging from a few \$100's million to billions of dollars. Rail projects usually consist of two main disciplines—civil and systems, with the civil component costing anywhere between 60 and 80% of the contract value. It is not unusual for rail projects, particularly transit/metro projects, to engage the services of technical specialist from a variety of technical disciplines such as architecture, landscape, fire and safety, roads, utilities, etc.

The integration of 'vertical' construction elements such as stations, parking facilities with 'horizontal' construction elements, such as track, bridges, and roadways, creates a need for a comprehensive set of design and construction services that is not normally found in other transportation projects. The nature and specialisation of these components usually requires two different entities to lead the design and construction efforts of each component. However, and in recent times, it is more common to award rail projects as a design and build contract in which the design and build contractor is a consortium comprising civil and system solution providers. Design and Build is a method of procurement in which a single legal entity takes full responsibility and sole liability for both design and construction [1]. The single legal entity may be a multi-disciplinary firm with in-house design capabilities or a consortium capable of providing a total solution. The design and build contractor is liable for all design and construction cost and must usually provide a firm fixed price in its proposal—these are typically lump sum contracts. In these type of projects, the design and build contractor commits to an aggressive schedule and implements strategies to enable the works to be completed to schedule and cost [2]. The design phase of any construction project is cyclic, repetitive and evolutionary involving designers from various design groups such as structural, mechanical, electrical and plumbing, architecture, road works. Often, these designers perceive their design scope with a unique and independent view neglecting the holistic view of the project. It is therefore not surprising that evidence exist that suggest that the design and construction failures originate from this ill-structured design process. It is therefore important that adequate effort must be taken to ensure a robust design strategy is in place from the onset and that all relevant stakeholders buy in the strategy [3]. One strategy implemented to reduce project delivery time is to reduce the design delivery time through the parallelism of sequential activities and it is not surprising therefore that many researchers have explored this aspect [4–9]. In this Chapter, a synopsis of the application of concurrent design principles and its applications to railway design and build projects is provided.

2. Engineering management

In a typical design and build project, the owner would have undertaken a 30% design effort. This design effort enables the owner to develop specific functional and performance requirements, establish preliminary stakeholder agreements, establish the alignment, secure land requirement, establish the capital cost estimate, minimise residual owner residual risk, etc. The owner's 30% design is usually supplied as part of the Request for Proposal (RFP)/Invitation To Tender (ITT) on an information basis with some components of the same, such as the alignment, supplied as owner's requirement. The design and build contractor is expected to complete the 70% design effort through a staged process that includes preliminary, detailed and final design. The completion of these design phases represents major milestones in the design life cycle and thus are typically referred to as design control points. The design and build contractor is expected to have performed a 100% design effort to complete the design delivery. To eliminate rework, it is preferable that design is complete or substantially complete before construction commences, thus an effective management of the design process is crucial to minimise cost and schedule overrun.

The design team is required to complete the design effort in earnest such that construction activities can proceed much earlier. This pressure on design has the objective of reducing the delivery time and minimising delivery cost. This task becomes more difficult as in most cases Systems design tends to follow a sequential progression of plans, specifications and products that are baselined and placed under configuration control. This sequential process, referred to as the Vee Model (also known as Verification and Validation Model), is usually specified in the Contract and mandated by International Standards such as IEC 62278 [10]. Furthermore, infrastructure owners are placing increasing emphasis on quality and reliability as well as the value proposition of the design and build contractor. The ability to deliver to schedule and cost is becoming a major differentiator in railway infrastructure projects.

The parallelism of sequential activities is in effect the application of concurrent engineering principles. There are numerous definitions for concurrent engineering, but the common theme in all such definitions is a holistic approach to product development that considers all life-cycle components and influences from the onset. For the purpose of this Chapter, the following definition by Cleetus [11] and Winner et al. [12] is preferred:

Concurrent engineering as a systematic approach to integrated product development that emphasises response to customer expectations and embodies team values of cooperation, trust and sharing in such a manner that decision making proceeds with large intervals of parallel working by all life-cycle perspectives early in the process.

Concurrent engineering is intended to ensure that contractors, from the onset of a railway infrastructure design-build project, consider all elements of the final system from conception through disposal, including quality, cost, schedule and user requirement [13, 14]. However, the overlapping of design activities may result in serious consequence if not managed effectively. Concurrent design is a holistic design approach that considers the constructibility of the product as part of the design and avoids design changes to enhances its constructibility.

2.1. Design management

The main objective in applying concurrent engineering to design is to reduce waste that may occur in the design cycle and to achieve continuous improvements in the design flow and output. This is achieved by viewing 'design' as [15]:

- A transformation of inputs to outputs;
- A process of information flow from one activity to another;
- A process of value generation.

Design is performed by a group of subject matter specialist whose main objective is the transformation of a client's requirements into outputs that comprise design decisions and actionable design documents. Tzortzopoulos and Formoso [16] identified three perspective of design:

- Conversion: In this view, the design is apportioned into sub-elements and assigned to a specialist who interpreted the client's requirements and converts the same into design decisions. Deshpande et al. [15] notes the tendency of occurrence of non-value adding components in the design when it is analysed simply as a conversion of inputs to outputs. Deshpande et al. postulated that such occurrence results in an increase in the time to complete design and/or insufficient time to generate optimal design solutions [17].

- Information Flow: Another school of thought, first proposed by Huovila et al. [18], suggest that the design process be viewed in terms of bi-directional information flow from stakeholders to the designers. A key principle of this thought is the identification and eradication of non-value adding activities from the design process.
- Value Generation: This school of thought on design is driven by the desire to achieve the best possible design outcome for the client. Huovila et al. [18] suggest that the process of value generation is dependent on the quality of information available to the designers, as well as the ability of the design team to transform complex, uncertain, and conflicting requirements into solutions that generate value for the client.

Ballard and Koskela [19] argued that it is necessary to integrate the three thoughts expressed above for effective design management. The quality of design can be improved by increasing the quantity and quality of available information with respect to customer needs and requirements. Requirements management in terms of apportionment, assessment, analysis and traceability is therefore a key component of design management. Tzortzopoulos and Formoso [16] provided practical guidelines for the implementation of lean concepts in the design process, these guidelines include:

- Identification and elimination of non-value adding activities in design;
- Increment of output value through detailed assessment of client requirements;
- Reduction of variability in the design process;
- Limiting the approval cycle times for design documents;
- Implementing design freeze and gate review concepts; and
- Establishing meaningful Key Performance Indicators (KPIs) and implementing continuous improvement in the design processes.

The design life cycle is typically separated into four stages – conceptual, preliminary, detailed and final design. Some projects specify a three-stage process consisting of preliminary, detailed and final design—thus for such projects, the initial design effort required represents a 60% design effort. Irrespective of the design life cycle, the Contractor is required, at the onset of the project, to assess and plan the works in terms of work breakdown structure that represents a detailed level at which appropriate reporting and earned value can be assessed [4]. The first step is to break down the design project into appropriate level of detail for budgeting and measuring progress. In this step, the work is broken down to level of details consistent with the requirements for scheduling and determining earned value. In most cases, an experienced rail design-build contractor will implement a breakdown, gained from experience on similar projects, based on an estimated number of design documents to be produced. The output of this first step, among others, is an estimate of the total quantity of design efforts in terms of configurable items (i.e. drawings, calculations, reports, software, specifications, etc) and identification of the design stages at which each configurable item will be delivered. Such a list is referred to as a Document Submittal Register (DSR) or as a Contract Data Requirement List (CDRL). It is acknowledged that the DSR is a live document that is updated throughout the life of the project as the design matures, however in reality attempts are made to freeze the DSR at final design.

The second step is to identify the design interfaces between the design work packages. These design interfaces determine the sequential dependency among design tasks. A matrix may be used to illustrate the dependencies between design work packages and between design tasks. In such a matrix, the columns represent predecessors awhile the rows represent successors. The matrix can be used to identify sequential relationships between design tasks. The third step is to separate the systems into independent groups. This involves grouping objects into homogenous groups, based on a set of common features. The goal of this process is to group dependent systems into manageable packages. The final step is to develop a network schedule, this may be represented using the precedence diagramming method or probabilistically using Programming Evaluation and Review Technique (PERT) [20]. A Graphical Evaluation and Review Technique (GERT) may be used to simulate and assess alternative branches of design activity loops [21].

A sequential design life cycle is illustrated in **Figure 1** below. In this life cycle, the development progresses through several defined phases. A detailed review of the differences and similarities between a sequential and concurrent logic is provided in [22].

This design logic is characterised by a sequential pattern where information about the product is slowly accumulated in consecutive stages. A stage commences only when the preceding

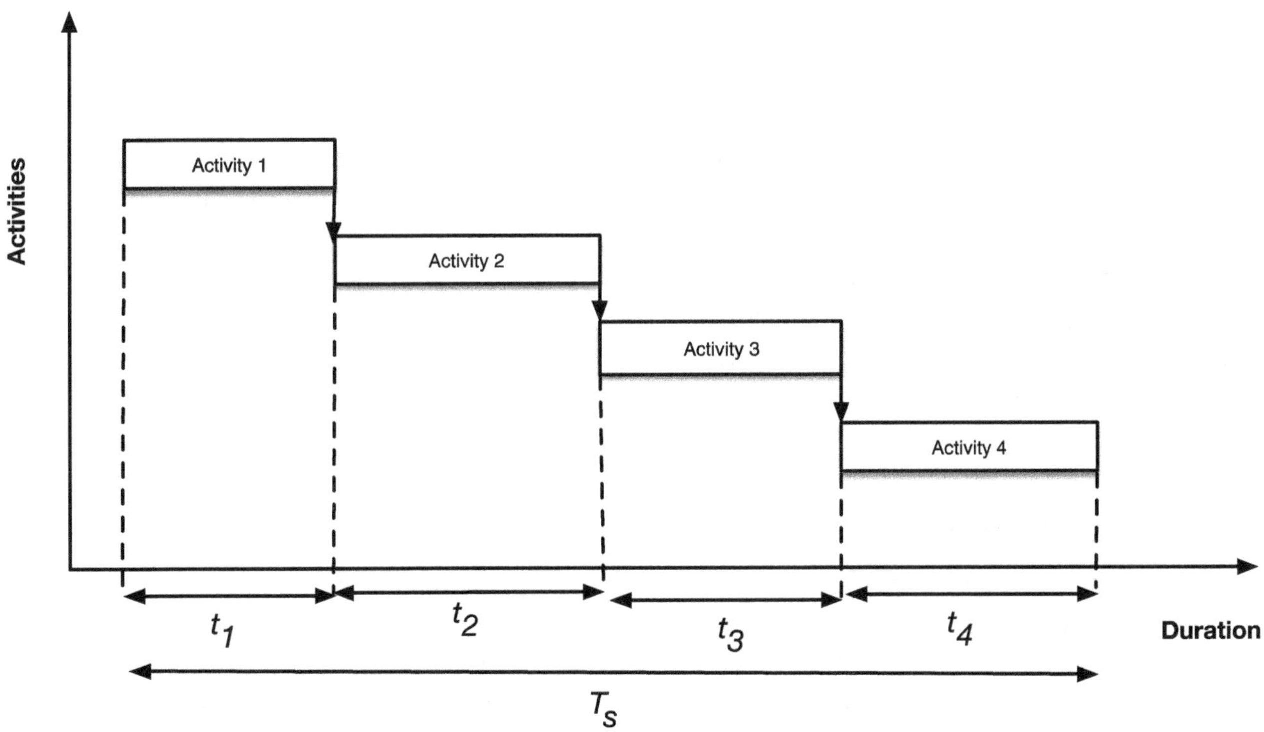

Figure 1. Sequential design logic.

stage is completed and has supplied complete and final information. This design life cycle is aligned with systems engineering principles and best suited for the system design component. Each stage must be completed before the next stage begins.

2.2. Concurrent engineering

Concurrent engineering involves reducing the total delivery time and cost of a project by overlapping activities (parallelisation of design activities) that are normally performed in a sequential manner. A core principle of the concept is the need to proceed at risk with an assumption that a specified performance will be obtained from a component, even before that performance has been demonstrated. The risk is managed on the basis of the integrity and certainty of the available information.

The extent of the overlap between two activities depends on the nature of the information exchange between these activities because it is the exchange of information that determines what work can start on the downstream activity. The extent to which two activities can be effectively overlapped depends on the relationship between those activities [22, 23]. Prasad [22] identified four types of relationships that are possible between design activities: (1) dependent activities, (2) semi-independent activities, (3) independent activities, and (4) interdependent activities. For dependent activities, the commencement of a downstream activity is dependent on the receipt of information from an upstream activity. Semi-independent design relationships commence upon the receipt of partial information from other activities. Independent design relationships are characterised by those activities that require no information from another before another activity can start. Interdependent design relationships are characterised by a bi-directional exchange of information between activities before either can be completed [24].

With respect to the identified design relationship types, only independent design activities can be overlapped without the risk of incurring delay or rework. There is an inherent risk in the overlapping of dependent activities. This inherent risk is due to the fact that a downstream activity, in a dependent activity relationship, must commence before all necessary information is available from a upstream activity. Thus changes in the upstream activity that impact assumptions made at the commencement of the downstream activity may increase the severity of the risk of delay or rework. This risk can be mitigated, in part, through increased communication and exchange of preliminary information between the upstream and downstream design activities. In other words, it is preferable not to concurrently design systems that belong to the same package.

2.2.1. Concurrent engineering characterisation

One way that concurrent engineering characterises the exchange of information is through the concept of information evolution of the upstream activity and the sensitivity of the downstream activity or activities to that information evolution [25]. Information and knowledge in an upstream activity can develop rapidly or slowly. For the downstream activity, the sensitivity to changes in the upstream information can be significant depending on the level of rework required. **Figure 2** illustrates the concept of a concurrent design logic. While applying

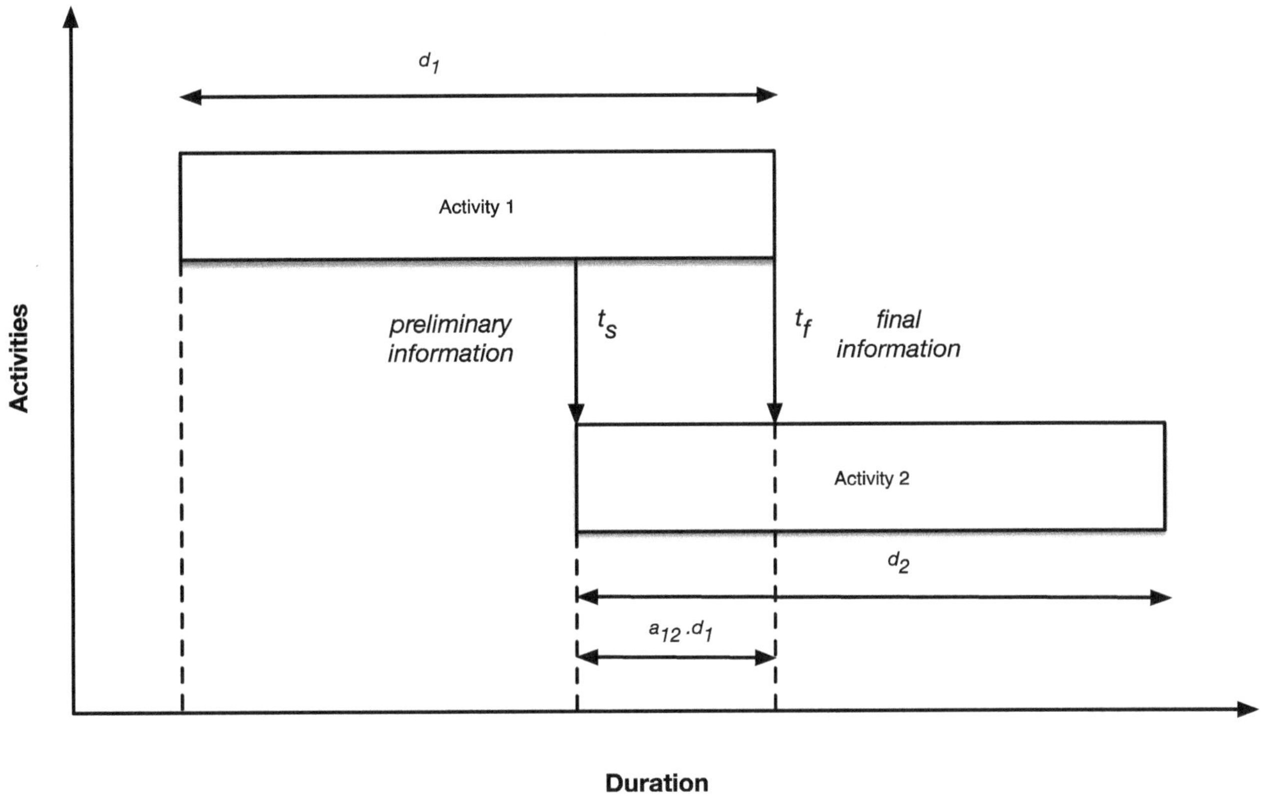

Figure 2. Concurrent design logic.

concurrent design logic, there are a number of conditions that need to be taken into consideration; the information exchange between a potential overlapping pair of activities, the management strategy used to facilitate the overlap between the pair(s), the likelihood of rework relative to the degree of overlap, and the impact of the rework on cost and schedule. These factors are analysed in the proceeding sections.

Concurrent engineering can be viewed as comprised of three basic components:

1. Simultaneity of Activities: In a sequential design flow, the total time, T_s, required to complete the design activity is given as:

$$T_s = \sum_{i=1}^{n} t_i \tag{1}$$

In the case of simultaneity of activities, the total time, T_c, required to complete the design activity is equal to the time duration of the activity with the maximum time duration:

$$T_c = \max(t_i)|i = 1, \ldots, n \tag{2}$$

Figure 3 illustrates the time required to execute the design activities in a concurrent design logic. Comparing the time required to complete the design activity in sequential design logic with a concurrent design logic, it is clear that a concurrent design logic offers a time saving of $\Delta T = T_s - T_c$

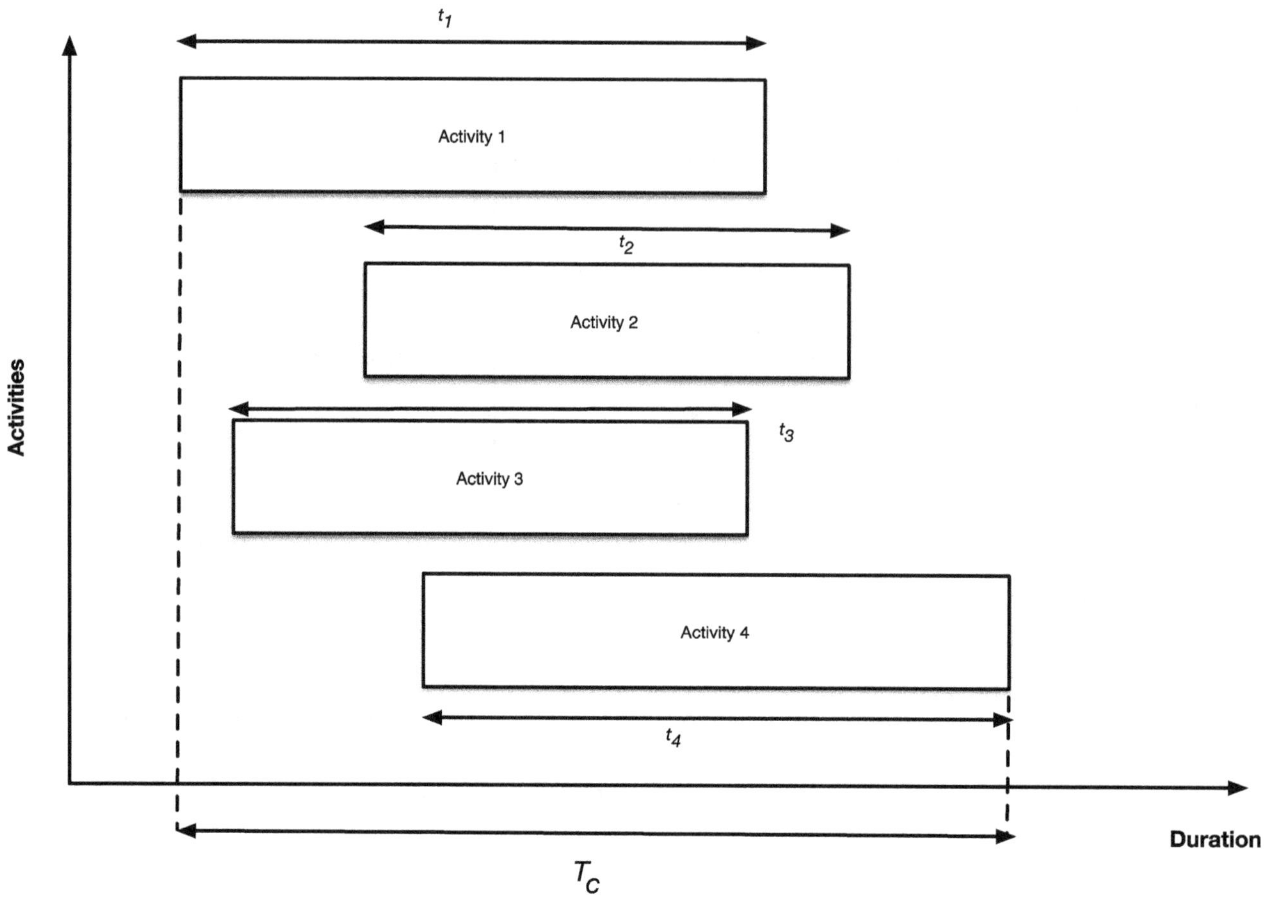

Figure 3. Time saving due to concurrent design logic.

2. Concurrency: The multifunctional design teams implement design concurrently (i.e. Activities 1, 2 and 3 are performed concurrently by different design teams) and interactively make decisions on works. Simultaneity of design activities without dynamic interaction of the various design teams does not assure concurrency. For example, consider a case of a simple overlapping of design activities in which communication is an acknowledgement of the conditions for commencing a task and those that underpin its completion. True concurrency of design implies interaction between the two activities in order to obtain the best decision, i.e. the two design activities 'concur' simultaneously for the best decision through dynamic interactions (communication), or solutions.

3. Simultaneity and concurrency need to occur at the onset or in the early stages of design process to ensure effective implementation of a concurrent design logic.

3. Factors impacting concurrent design logic

The greatest impact and benefits of concurrent engineering is evident at the design stage. The design decisions made in the early design stages (i.e. conceptual and preliminary design phases) have a significant impact on the constructibility of a product, as between 70 and 80% of the construction cost is determined by design [26, 27]. Thus, cost reduction efforts must be an integral component of the design effort. In the following, we review factors that need to be considered to achieve a successful concurrent design output.

3.1. Information evolution

As previously mentioned, concurrent design logic can be viewed as an information processing system in which individual design activities are modelled as information processing units that receive information from proceeding activities and transform the information received into new information to be passed on to subsequent activities. With reference to **Figure 2**, preliminary information of the design Activity 1 is available t_s and is continuously modified until the end of the activity. Activity 2 can start at any time between t_s and t_f. Evolution describes the rate at which design information is generated from the start of an activity through the completion of the activity. It is acknowledged that in practise a quantitative assessment of information evolution may be impracticable and thus a qualitative approach is favoured. There are four key determinants of evolution:

- Design optimisation: The level of optimisation performed on design elements or the number of design alternatives evaluated
- Constrain satisfaction: The flexibility of design elements in satisfying constraints
- External information exchange: The amount of information received from or reviewed by external sources; and
- Standardisation: The level of standardisation in the design process and/or design product

Each of the determinants of evolution listed above relies on activity iteration as a determining factor. Design information in those activities with iteration evolves slower than activities without iteration. It goes without saying that an activity without constraint or pressure will evolve naturally and that this natural evolution tends to produce the best design outcome for that activity, however, most design is performed under some constraint, and this is particularly true under a design and build project. Such constraints results in actions that alter the natural evolution of an activity; for example, actions resulting from time constraint may results in the reduction of the time taken to complete an individual activity or a reduction in the overall design schedule.

However, the gains from overlapping must be balanced against the potential of rework (cost and time) which results from the modification of the upstream information. When preliminary upstream information is utilised by the downstream activity too early, future changes may have to be incorporated in time consuming subsequent iterations that result in an increase downstream duration and effort. The amount of rework required, if preliminary information changes, is a function of the sensitivity of the downstream activity to changes in the upstream information.

3.2. Sensitivity

Krishnan et al. [28] qualified sensitivity as a measure of the amount of rework required in a downstream activity as a result of information evolution in an upstream activity. The following conditions impact the sensitivity severity of the downstream activities to changes in upstream information and thus increase the risk of rework:

- The downstream design is near a constraint or boundary;
- The downstream design depends on one key upstream input; or
- The downstream design is integrated such that changes cannot be isolated.

Small changes in the upstream information could result in extensive rework with a major cost and schedule impact to a highly sensitive downstream activity. On the contrary, a low sensitivity downstream activity can accommodate changes in information from an upstream activity such that minimum or no rework is required with minimal cost and/or schedule impact. Bogus and Molenaar [29] identified three determinant factors that influence the sensitivity of an activity:

- Constraint sensitive: The proximity of the downstream design to a constraint or boundary;
- Input sensitive: The level of dependence of downstream design on specific inputs from other activities; and
- Integration sensitive: The ability of the downstream design element to be separated from the entire system.

The combination of an upstream activity with a fast or slow evolution and a downstream activity with a low or high sensitivity results in four possible combinations of evolution and sensitivity. These four possible combinations are major considerations in the assessment of the probability of rework for an activity pair. Roemer et al. [8] and Bogus and Molenaar [29] defined rework as the *"increase in time and costs, direct and indirect, that are required to correct some of the work in the downstream activity due to incorrect or changing information received from the upstream activity"*. This definition highlights the importance of the need to ensure the integrity of the underpinning assumptions and information flow from the upstream activity.

4. Risk mitigations

The need to commence railway construction activities in earnest serves to meet the aggressive schedule imposed through the contract. These projects, typically structured under International Federation of Consulting Engineers (FIDIC) rules, place the Contractor as the majority owner of associated risk. The design and build contractor therefore needs to put in place adequate processes and control to manage the delivery and in particular the cost. It is beneficial to apply the principles of concurrent design at the commencement of the project, with due consideration of requirement management, design freeze, over-design, etc.

4.1. Design freeze

Eger et al. [30] defines design freeze as a *"binding decision that defines the whole product, its parts or parameters and allows the continuation of the design based on that decision"*. Design freeze allows structuring and planning of the design process [30]. Freezing a design or key components of a design aims to reduce the likelihood of engineering changes, however, any change required to be implemented after a design freeze may result in high rework cost and potential delays. Design freeze can apply to different stages of the design life cycle. **Figure 4** shows a typical design gate review process; it is easily recognised that the logic shown in **Figure 4** is a sequential logic, however in reality it is possible to apply design freeze in a concurrent design process to facilitate the early commencement of a downstream activity, however, depending on how it is implemented, design freeze in a concurrent design logic can be viewed as performing the design activities in sequential manner using incomplete preliminary information from upstream activities. In this case, the risk of possible design changes increases with greater degrees of overlap. There are many advantages of the application of the concept of design freeze; it can facilitate the early procurement of long lead items; it can also assist in the reduction of the risk of rework and can set preliminary information from an upstream activity as a basis for further work. Once design freeze has occurred, changes to downstream activities resulting from evolution of preliminary information of the upstream activities needs to be carefully analysed before proceeding. Alternative implementation strategies should be considered and all changes should follow a change control process.

4.2. Overdesign

Unlike design freeze, overdesign adds a margin of safety to the design as an attempt to mitigate potential errors in the information flow during overlapping periods. It can be defined as the process of implementing conservative assumptions, in the downstream activity in lieu of incomplete preliminary information transfer from the upstream activity. As an example, one may make conservative assumptions on the required size of technical rooms, while the systems design is still in its infancy, with the view to allow construction of a station or depot to proceed. There is, however, an inherent risk that the margin of safety applied might not be adequate and thus resulting in an underdesign scenario. This may result if the initial assumption is based on previous project experience without adequate analysis and resolution of the current contract's

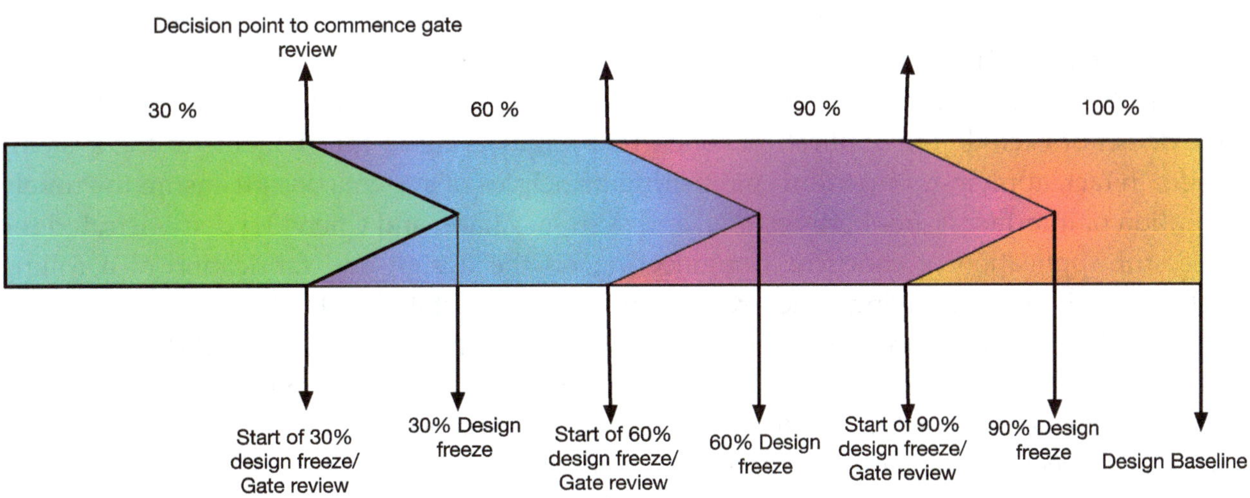

Figure 4. Design gate review.

requirements, particularly those concerning the civil-system interfaces. Such a scenario may result in rework with cost and time impact. There is therefore a balance to be maintained between the robustness and integrity of the underpinning design assumptions and the cost of implementing the design. A trade-off also exist between the degree of overlapping and the certainty of upstream information.

4.3. Standardisation

Standardisation is the process of adopting a design solution to be used repetitively on a project. Such practice speeds up the evolution of upstream activities and enables early information transfer from the upstream activity to the downstream. There is a likelihood of cost increases due to lack of design optimisation. In recent applications of this technique on a design and build project with 22 stations comprising of elevated and underground stations, two archetypes stations representing an elevated and underground stations where processed through design completion and these archetypes acted as the design standards for the remainder of the stations designs. This approach resulted in increased construction productivity. It should be noted that subtle difference between stations—in terms of size, layout etc. —may require additional design effort over and above that established in the standard design. It is to be recognised that standardisation in terms of design processes and procedures further contributes to an increase performance of the construction output and the overall project schedule.

5. Systems design

As mentioned in Section 2.1, the tendency is for Systems design to follow a sequential design logic, partly because of the safety-critical nature that rail systems serves within the railway infrastructure and partly because the systems assurance process forces a sequential logic review. That said, in most design and build projects, the initial delay manifest from the civil design and construction phases. Systems, being the last major component of the Works, therefore are under constant pressure to mitigate the delays incurred from a predominate civil upstream activity. The systems activities under such pressure tend to be Systems installation and Test & Commissioning. The fact that systems design tends to follow a sequential design logic does not exclude the application concurrent engineering to the elements of the systems design. In fact, applying concurrent engineering principles of systems design ensure the timely resolution of interface issues between civil and system. Zhang and Chen [31] demonstrated the successful application of concurrent engineering on the design and fabrication of a rolling stock. The driver for applying concurrent engineering was stated as to shorten product engineering delay, improve locomotive design and capitalise knowledge. Park [32] demonstrated that concurrent design principles can be applied to safety-critical system using a model-based approach. Furthermore, while IEC 62278 implies a sequential logic, ISO/IEC 24748 [33] emphasises that projects should integrate the concurrent design of products and their related lifecycle processes. It goes on to state that *'concurrent engineering should integrate product and process development to ensure that the product(s) are producible, usable, and supportable'*.

6. Conclusion

This Chapter discusses the application of concurrent engineering concept and principles to the design process of a design-build rail project. It is identified that concurrent engineering is a logical approach to achieve a reduction in project delivery time and cost. It is highlighted that the key objective of meeting the desired project duration and cost expectations is through the overlapping of dependent activities. It is noted that overlapping should be approached in a systematic manner to reduce costs and risks. While concurrent engineering is not a term typically associated with design and build rail project, the concept is not alien to the rail construction industry as attempts at mitigating delays, avoiding potential delay penalties and cost overrun due to retrofits and delays always results in an 'accelerated' schedule which typically exhibits the application of concurrent engineering logic in what was otherwise a sequential logic. It is highlighted that executing the design activities of a railway design and build project concurrently will result in improvements in quality, time to deliver, cash flow and profitability, etc. It is crucial that the designers, schedulers and planners work together from the onset to develop the project schedule reflective of concurrent engineering logic.

Author details

Ade Ogunsola

Address all correspondence to: ade.ogunsola@mac.com

Parsons Infrastructure Group, Riyadh, Saudi Arabia

References

[1] Anumba CJ, Evbuomwan NFO. Concurrent engineering in design-build projects. Journal of Construction Management and Economics. 1997;**15**(3):271-281

[2] Law C-YE. The Application of Concurrent Engineering in the Construction Process in Hong Kong. University of Hong Kong; Hong Kong. China; 2002

[3] Mujumdar P, Maheswari JU. Design iteration in construction projects – Review and directions. Alexandria Engineering Journal. January 2017:1-9. https://doi.org/10.1016/j.aej.2016.12.004

[4] Dzeng RJ. Identifying a design management package to support concurrent design in building wafer fabrication facilities. Journal of Construction Engineering and Management. 2006;**132**(6):606-614

[5] Gurevich G, Keren B, Laslo Z. The problem of overlapping project activities with interdependency. International Scientific Journal: Science, Business and Society. September 2016;**1**(6):18-21

[6] Grèze L, Pellerin R, Leclaire P, Perrier N. Evaluating the effectiveness of task overlapping as a risk response strategy in engineering projects. International Journal of Project Organisation and Management. 2014;**6**(1/2):33-16

[7] Lin J. Overlapping in distributed product development. In: 2006 International System Dynamics Conference. System Dynamics Society. Nijmegen, The Netherlands; 2006

[8] Roemer TA, Ahmadi R, Wang RH. Time-cost trade-offs in overlapped product development. Operations Research. 2000;**48**(6):858-865

[9] Terwiesch C, Loch CH. Measuring the effectiveness of overlapping development activities. Management Science. April 1999;**45**(4):455-465

[10] IEC 62278. Railway Application-Specification Administration of Reliability, Availability, Maintainability and Safety. Technical report. IEC, Geneva, Switzerland; 2015

[11] Cleetus KJ. Definition of Concurrent Engineering. CERC (Concurrent Engineering Research Center) Technical Report Series, Research Notes Technical report, CERC-TR-92003. West Virginia University; USA, 1992

[12] Winner RI, Pennell JP, Bertrand HE, Slusarczuk MM. The Role of Concurrent Engineering in Weapons System Acquisition Technical report. Alexandria, VA, USA: Institute for Defence Analyses; 1988

[13] Minnaar S, Reinecke R. Concurrent engineering with constraint networks. The South African Journal of Industrial Engineering. 1993;**7**(2):1-11

[14] Pennell JP, Winner RI. Concurrent engineering: Practices and prospects. In: IEEE Global Telecommunications Conference, 1989, and Exhibition. Communications Technology for the 1990s and Beyond. IEEE. Dallas, USA. pp. 647-655

[15] Deshpande AS, Filson LE, Salem OM. Lean techniques in the management of the design of an industrial project. Journal of Management in Engineering. April 2012;**28**(2):221-223

[16] Tzortzopoulos P, Formoso C. Considerations on application of lean construction principles to design management. In: Proceedings for the 7th Annual Conference of the International Group for Lean Construction (IGLC). 1999: 335-344

[17] Tzortzopoulos P. Contribuições para o desenvolvimento de um modelo do processo de projeto de edificações em empresas construtoras incorporadoras de pequeno porte [Master's thesis]. Porto Alegre-RS, Brazil: Federal Univ. of Rio Grande do Sul; 1999

[18] Huovila P, Koskela L, Lautanala M. Fast or concurrent: The art of getting construction improved. In: Proceeding of the 2nd International Conference on Lean Construction. Vol. 143. Rotterdam, The Netherlands: AA Balkema; September 1997. p. 149-166

[19] Ballard G, Koskela L. On the agenda of design management research. In: Proceedings for the 6th Annual Conference of the International Group for Lean Construction (IGLC). Guarujá, Brazil. 1998

[20] Ahuja HN, Dozzi SP, Abourizk SM. Project Management: Technique in Planning and Controlling Construction Projects. Wiley, USA; 1994

[21] Nelson RG, Azaron A, Aref S. The use of a GERT based method to model concurrent product development processes. European Journal of Operational Research. April 2016; **250**(2):566-578

[22] Prasad B. Concurrent Engineering Fundamentals, Vol I: Integrated Product and Process Organization. 1st Edition: 1996. Prentice-Hall international series in industrial and systems engineering. Prentice Hall PTR, Upper Saddle River, NJ, USA

[23] Yassine AA, Chelst KR, Falkenburg DR. A decision analytic framework for evaluating concurrent engineering. IEEE Transactions on Engineering Management. May 1999;**46**(2):144-157

[24] Bogus SM, Diekmann JE, Molenaar KR. Simulation of overlapping design activities in concurrent engineering. Journal of Construction Engineering and Management. 2011;**137**(11): 950-957

[25] Krishnan V, Eppinger SD. A model-based framework to overlap product development activities. Management. 1997;**43**(4):437-451

[26] Corbett J, Crookall JR. Design for economic manufacture. CIRP Annals-Manufacturing Technology. 1986;**35**:93-97

[27] Mileham AR, Currie GC, Miles AW, Bradford DT. A parametric approach to cost estimating at the conceptual stage of design. Journal of Engineering Design. March 2007;**4**(2):117-125

[28] Krishnan V, Eppinger SD, Whitney DE. Accelerating product development by the exchange of preliminary product design information. Journal of Mechanical Design. December 1995; **117**(4):491-498

[29] Bogus SM, Molenaar KR. Concurrent engineering approach to reducing design delivery time. Journal of Construction Engineering and Management. 2005;**131**(11):1179-1185

[30] Eger T, Eckert C, Clarkson PJ. The role of design freeze in product development. In: DS 35: Proceedings ICED 05, the 15th International Conference on Engineering Design, Melbourne, Australia. 2005: 1-11

[31] Zhang H, Chen D. Developing a multidisciplinary approach for concurrent engineering and collaborative design. In: The 8th International Conference on Computer Supported Cooperative Work in Design. IEEE, Xiamen, China. 2004. p. 449-454

[32] Park JY. Model-based concurrent systems design for safety. Concurrent Engineering. December 2004;**12**(4):287-294

[33] ISO/IEC/IEEE 24748-4. International Standard for Systems and Software Engineering – Life Cycle Management – Part 4: Systems Engineering Planning. IEEE, USA; 2016

7

Model-Based Fault Analysis for Railway Traction Systems

Jon del Olmo, Fernando Garramiola, Javier Poza and Gaizka Almandoz

Abstract

Fault analysis in industrial equipment has been usually performed using classical techniques such as failure modes and effects analysis (FMEA) and fault tree analysis (FTA). Model-based fault analysis has been used during the last several years in order to overcome the limitations of classical methods when complex industrial equipment has to be analyzed. In railway and automotive sectors, the development and validation of new products are based on hardware-in-the-loop (HIL) platforms. In this chapter, a methodology to enhance classical FMEAs is presented. Based on HIL simulations, the objective is to improve the results of the fault analysis with quantitative information about the effects of each fault mode. In this way, the impact of the fault analysis in the design of the traction system, the development of new diagnostic functionalities and in the maintenance tasks will increase.

Keywords: railway, traction, fault, models, FMEA

1. Introduction

Nowadays, reliability, availability, maintainability and safety (RAMS) are key features in the development of industrial equipment. Moreover, new maintenance approaches, such as condition-based maintenance (CBM) have emerged, as an alternative to preventive maintenance and run failure techniques [1]. In sectors such as railways, automotive and aviation, the business model is based on the sale of equipment and its long-term maintenance. Taking into account that locomotives might have a service life of up to 30 years, a long interval of the life cycle is related to maintenance tasks and technical services. This has been an incentive for manufacturers to improve the reliability and maintainability of their products [2]. During the last years, several projects were launched in order to develop smart maintenance systems [3–6].

From the very first phases of the Life Cycle, reliability and safety analysis are performed. In these analyses, the effect of faults in the functionalities of the system is studied. In this field, techniques such as FMEA and FTA allow the designer to identify systematically fault modes (FMs) and effects. However, lately some limitations have been identified in the application of these classical methodologies, especially when a complex control system has to be analyzed. Moreover, more and more manufacturers are using Model-based techniques [7] in the development of new systems, following the global tendency of Model-based engineering. Fault analysis is not an exception and several authors have proposed to use models to analyze the effects of faults [8–10].

The extensive use of models for the development of railway systems has allowed the introduction of tools such as hardware-in-the-loop (HIL) platforms. These platforms have a key role in the validation of embedded control units, just as they have in automotive applications. Thanks to this kind of platforms, development time and costs are reduced considerably.

Taking into account the limitations of classical fault analysis methods and the extensive use of HIL platforms in the railway traction system manufacturing sector, in this chapter, a methodology for model-based fault analysis that takes advantage of HIL platforms is presented. Concretely, this methodology has been developed to improve the analysis of faults and effects of the railway traction system shown in **Figure 1**. This traction converter box has a three-phase inverter supplying two induction motors in parallel. Moreover, it has a breaking chopper, a DC-Link, an input filter and voltage, current and speed sensors. The control of the converter is executed by an embedded traction control unit (TCU).

The main goal of the methodology is to quantify the effects of the faults using analytical models and simulation platforms. Generally, the conclusions of an FMEA are used to improve designs, in future redesigns, identify new maintenance tasks and develop new fault detection and identification algorithms. With model-based approaches, the quality of the analysis improves and its impact on the three aforementioned aspects is increased.

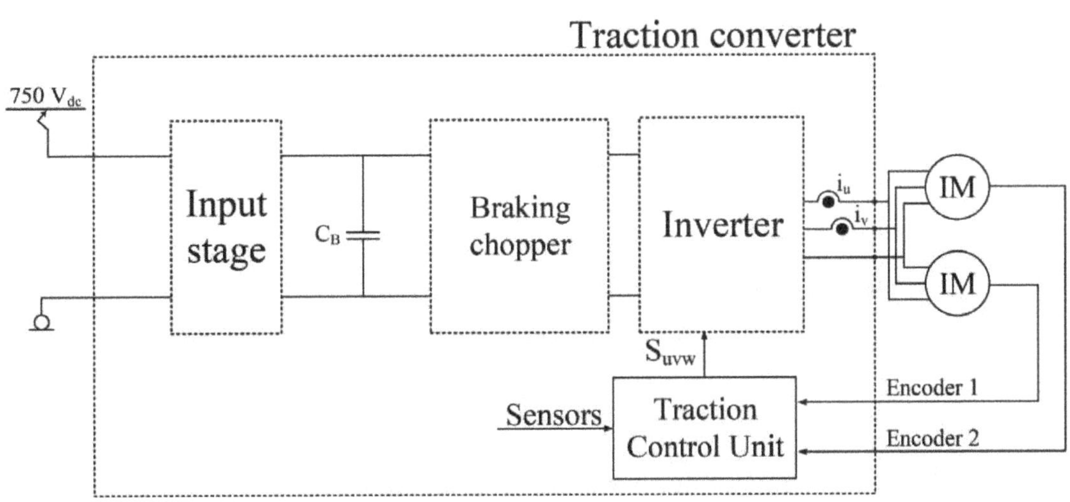

Figure 1. Tramway traction system schematics.

The chapter is organized as follows. In Section 2, classical fault analysis methods and their limitations are described. In Section 3, a brief state-of-the-art about model-based fault analysis is presented. In Section 4, the new model and HIL-based methodology is shown, accompanied by a case study. Finally, in Section 5, the conclusions are drawn.

2. Fault analysis in complex control systems: classical methods and limitations

2.1. Classical faults and effects analysis methods

In this section, a brief description of classical fault analysis methods will be presented. It provides the reader with a basic understanding of the most used techniques in the analysis of faults and their effects in complex control systems.

2.1.1. Failure modes and effects analysis (FMEA)

Fault modes and effects analysis (FMEA) is an inductive method used in the development of products in order to identify and classify fault modes and effects. This technique establishes a systematic approach to identify effects for each fault mode and classify them in terms of occurrence of probability and severity. Occurrence and detection probabilities are combined with severity indexes to obtain a risk priority number (RPN). Hence, corrective actions can be prioritized [11]. Recently, due to the deficiencies of RPN application in real-world cases, enhanced approaches have been proposed [12].

The main steps to perform the FMEA analysis are [11]:

Before the analysis:

1. Gather information about requirements, components, architecture and fault modes.
2. Organize the system under analysis in a structured way. Describe the architecture and its limits in a block diagram.

During the analysis:

3. Fault mode analysis and FMEA worksheets:
    a. Determine fault modes.
    b. Determine the effects of each fault mode.
    c. Determine the causes of each fault mode.
    d. Determine the existing protection actions.
    e. Obtain the RPN of each fault mode.
    f. Prepare FMEA tables.

After the analysis is completed:

4. Summarize the analysis in an FMEA report.
5. Establish corrective actions to mitigate the effects of the faults.

At the end of this process, an FMEA worksheet is obtained. Several worksheet examples can be found in the literature [11, 13].

Table 1 shows an excerpt from an FMEA worksheet of a railway traction system. In particular, the behavior of a railway traction drive under phase current sensor faults is analyzed.

2.1.2. Fault tree analysis (FTA)

A fault tree is a graphical representation of all the basic events that can cause an undesired event in a process or system. The faults may be related to hardware components, human errors or any other event that can generate an undesired situation. Therefore, a fault tree describes the logical relation between basic events that can lead the system to a faulty state.

It is important to understand that a fault tree is created for its main undesired event. Hence, it does not describe all the faults that can occur in a system and more than one Fault Tree is needed to describe the faulty behavior. **Figure 2** shows an example of a fault tree and the steps to build it.

The analysis performed with the FTA technique consists of the following steps:

1. Define the undesired event under study. A unique fault tree is obtained for each event.
2. Understand the system. All the causes that can lead to the main event are analyzed.
3. Fault tree construction. Events and causes are linked using logic gates. This step is an iterative process. An event is selected and its causes are identified. These causes are classified as basic events, undeveloped events or intermediate events and the logical gates

Fault mode	Cause	Local level effect	Traction unit level effect	Train level effect
Measured value bigger than real value	Internal failure	False measurement	Inappropriate control Overcurrent Disabled converter.	Loss of traction unit
No measurement	Internal failure	No measurement	Inappropriate control Overcurrent Disabled converter	Loss of traction unit
Open-circuit	Internal failure	No measurement	Inappropriate control Overcurrent Disabled converter	Loss of traction unit
Measured value smaller than real value	Internal failure	False measurement	Inappropriate control Overcurrent Disabled converter	Loss of traction unit

Table 1. Excerpt from an FMEA worksheet of railway traction systems: Phase current sensor FMEA.

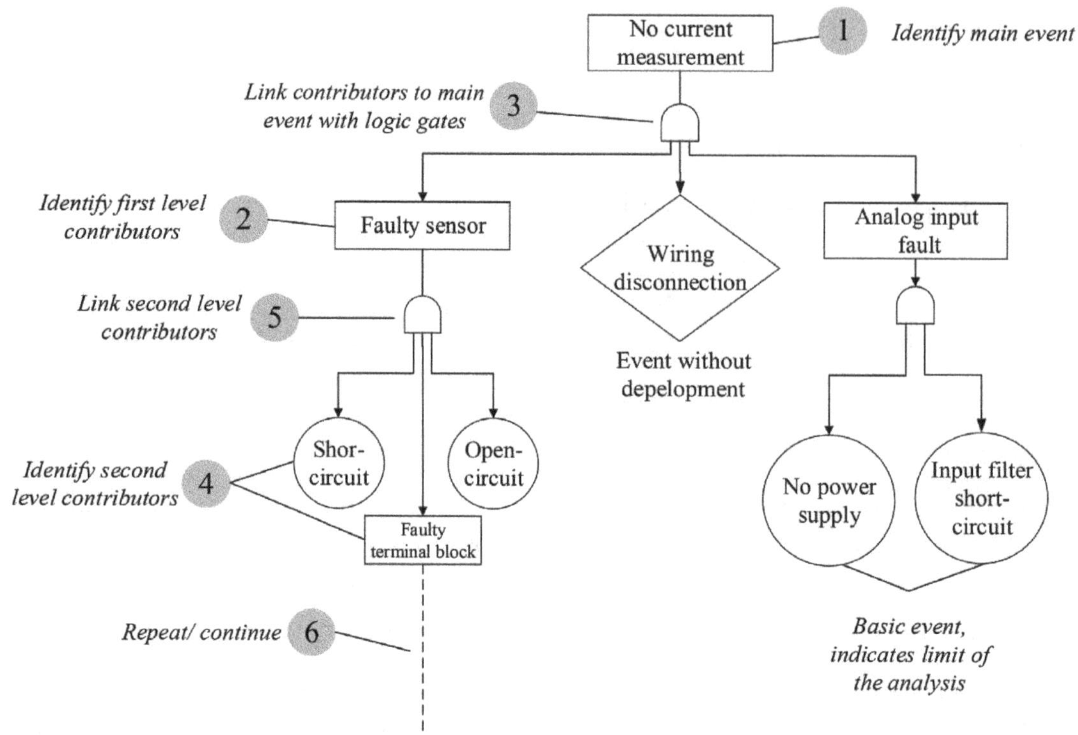

Figure 2. FTA for current sensor.

to link all of them are chosen. In this way, the tree is drawn from the top event to the basic events (see **Figure 2**).

4. Evaluate the fault tree. It is analyzed in order to suggest improvements. Moreover, the risk each fault generates is assessed.

5. Control measures are proposed. Once the risks associated with each fault or undesired event are identified mitigation measures are proposed.

2.2. Limitations

One of the challenges that railway traction system manufacturers have to face is the difficulty of combining and coordinating product development and safety analysis tasks. The integration of systems engineering and safety analysis is addressed in many other research works [14–16]. Nowadays, the fault analysis is mainly performed using the tools described in the previous section. During the design stage, the safety analysis is performed in order to achieve two main objectives. The first one is to draft a safety case document that allows the manufacturer to obtain the corresponding safety certificate. The second is to analyze the architecture of the system under development to ensure that meets availability, reliability and maintainability requirements. Nevertheless, as the system is analyzed from the high level (only the architecture is studied and implementation details are not taken into account), design teams rarely receive any feedback about the details of the effects and the potential improvements to mitigate those effects.

If the fault analysis is intended to be a source of information for future system improvements, redesigns and maintenance task, this analysis has to be more detailed and performed in collaboration between the design and the safety teams. Classical fault analysis methodologies lack tools to manage this need.

In addition, techniques such as FMEA and FTA are performed manually and they are based on requirement documents and informal models of the system [9]. With informal models, we refer to architecture models developed in the early stages of the design process. In the life cycle of a product, the safety analysis is part of the design step and it does not usually reflect the changes that the system has experienced later in the implementation and validation phases. Hence, these analyses are usually incomplete.

Apart from the aforementioned shortcomings, the following limitations have been also identified by other authors [8, 9, 17, 18]:

- They represent the fault logic in a static way and they do not allow analyzing neither the time information nor the dynamic behavior of the system. For instance, in railway traction applications a conventional FMEA does not usually reflect the behavior of the control strategy under faulty conditions. It is difficult to take into account the following aspects using classical methods: the implementation details of the control strategy, the transitions between control modes, the changes between operation points and the interaction between different subsystems when a fault occurs.
- The analysis depends on the skills of the analysist to predict the effects of each fault. Apart from the difficulty to reflect the dynamic behavior with a classical method, the analyst needs to know the specifics of the control strategy and its implementation. It is worth mentioning that under faulty conditions, a closed loop control reacts to compensate the effects of the fault. This is an added difficulty in the fault analysis process since fault effects are modified by the control system itself.
- Classical methods do not provide with efficient tools to manage the complexity and the number of components of current industrial equipment.

3. State of the art of model-based fault modes and effects analysis techniques

3.1. Introduction

Model-based safety analysis has been developed during the last years to help the analysis of complex systems, taking as a central element the model and automating the analysis of extended fault models [19]. This new approach intends to overcome the limitations mentioned in Section 2.2.

Model-based safety analysis methodologies can be divided into two main groups: failure logic modeling (FLM) and behavioral fault simulation (BFS), also known as fault injection (FI) [8, 10].

In the following section, these two alternatives will be described and the selection of fault injection as the basis for the new methodology will be explained.

3.2. Model-based fault modes and effects analysis techniques

3.2.1. Failure logic modeling approach

This methodology is based on the automatic generation of fault trees and FMEA worksheets using the information stored in models [8]. For each component of the model, its inputs, outputs and behavior under fault are defined [10, 20]. To specify the behavior, the following elements have to be described:

- Input fault modes.
- Internal fault modes.
- The logic that defines the effect of the input and internal fault modes in the output.

Figure 3 shows the definition of fault modes using this methodology.

Once the fault logic is defined for each component, a fault model can be developed linking the outputs of each block with the inputs of the next block. The components are connected as defined in the architecture, which allows reflecting the structure of the system in the posterior FMEA analysis. With this new model, the propagation of each fault mode can be studied. Moreover, the homogeneous and systematic description of each element (with output fault modes as a function of input fault modes and internal fault modes) allows automating the analysis process.

In the literature, several techniques and tools can be found to automate this process following the failure logic modeling approach [8]. Among others, from the railway traction application point of view, the HiP-HOPS methodology and its associated tools are the most interesting ones [19, 22]. This tool, implemented as a Matlab/Simulink tool, allows the analyst to define

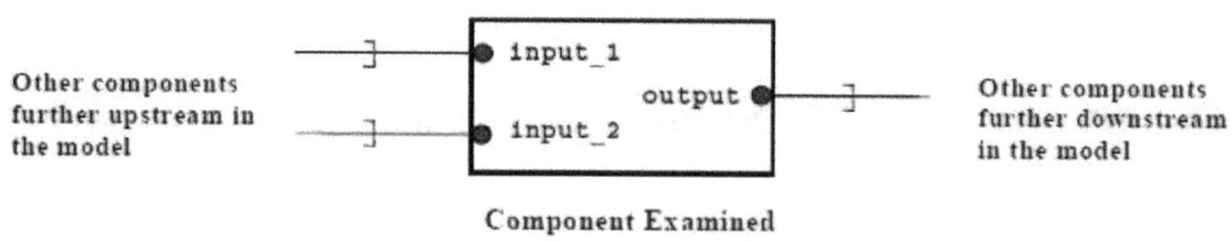

Output Failure Mode	Description	Input Deviation Logic	Component Malfunction Logic	λ(f/h)
Omission-output	The component fails to generate the output	Omission-input_1 AND Omission-input_2	Jammed OR Short_circuited	$5x10^{-7}$ $6x10^{-6}$
Wrong-output	The component generates wrong output	Wrong-input_1 OR Wrong-input 2	Biased	$6x10^{-8}$
Early-output	Early output	...	...	...

Figure 3. Definition of fault modes for fault analysis using FLM and hip-HOPS methodology [21].

the fault behavior of the system over Simulink models. This is an advantage in the field of electric drives because Matlab/Simulink is one of the main tools used for their development.

With FLM methodologies, some of the limitations mentioned before are overcome. The efforts have been directed towards the automation of the analysis and characterization of faulty systems using extended models. Nevertheless, as it was mentioned in [18], the main limitation that FLM techniques have to face is the lack of tools to analyze the dynamic behavior of the system. They do not consider the changes in the states of a system and are not able to model its dynamic or temporal behavior [19]. A common framework to present the structure and analyze the propagation of the system is defined, but the information about the dynamic behavior of the system (operation modes, reaction to faults, etc.) is not used. In this respect, some authors have proposed alternatives that take into account the dynamic behavior of the system. In the study by Kabir [23] one of the main objectives is the modeling of the behavior of the system using state machines.

3.2.2. Behavioral fault simulation/failure injection approach

The BFS technique is based on the injection of faults using executable models of the system to define their effects [8]. The starting point for the analysis is a formal or nominal model (without faults) known and used during the design stage (see **Figure 4**). This model is extended with information related to the faults of the system. In this way, the effects of the faults and the behavior of the system when the faults occur can be analyzed. The extended and common model assures that the results of the fault analysis are relevant and that are updated with respect to design changes.

The key to this approach is that the models are executable and they allow analyzing the dynamic behavior of the system under faults. An analysis platform is used to simulate the extended models and assure that the system meets safety requirements.

Up to now, the work related to fault injection has been focused on the development of simulation platforms to apply the described methodology. In [24] and [25] the results of two European projects were presented, where this technique was applied to models developed in

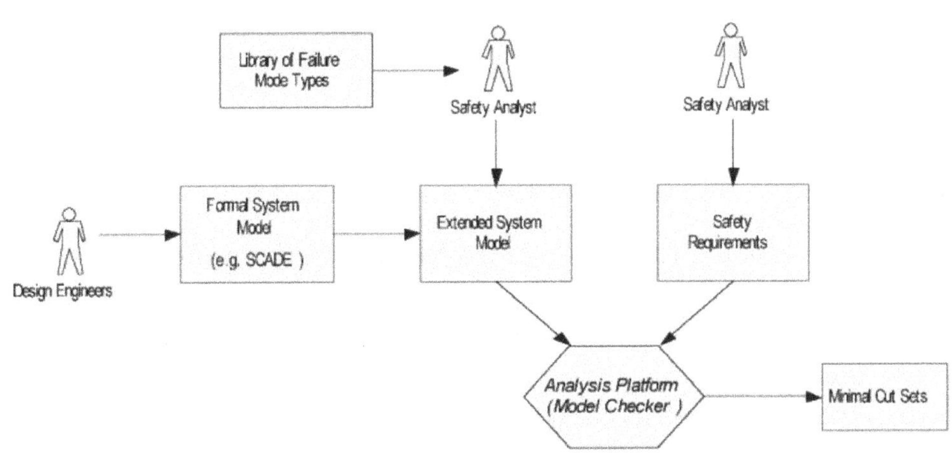

Figure 4. Principal components of failure injection approach [10].

SCADE and Statement software. In [9], the nominal models developed in Simulink were translated into SCADE that has a module for fault mode analysis.

Compared to FLM methods, BFS approaches allow simulating the dynamic behavior of the system. Moreover, as the models already validated in the design phase are the basis for the analysis, the results obtained about the effects are more accurate. Hence, the fault analysis does not entirely depend on the knowledge of the analysist because the system and its behavior have been defined in the model. In any case, some disadvantages have been also identified. First, since extended dynamic models are used, there is an inherent difficulty to process automation, as well as, not to generate an excessive amount of models. It has to be taken into account that in complex systems there could be a lot of different fault modes. Each fault mode could demand a different model and simulation in order to perform the fault analysis. In some systems, the number of simulations could be unmanageable. Second, another drawback of BFS methods is that it can be only applied after the designs and the models are developed, in the later stages of the development process [10]. This makes the introduction of changes difficult if the system does not comply with requirements.

However, it is worth mentioning that nowadays many manufacturers have chosen model-based systems engineering [26] as their main approach for the development of complex systems, avoiding document-based product development. This methodology is based on the extensive use of models during the whole life cycle of a product, from the requirement definition phase to the validation stage. Therefore, this context is ideal for the deployment of model-based fault injection methods for fault analysis.

4. Hardware-in-the-loop-based FMEA for railway traction applications

4.1. Need for a quantitative fault analysis methodology

As it was mentioned in the introduction, this work is related to the development of electric drives for railway traction applications. The results obtained using the classical fault analysis methods were assessed using reference information provided by the manufacturer CAF Power & Automation. The limitations identified in the literature were also found in the industrial application.

Therefore, the objective of the work presented here was to propose a new fault analysis methodology in the field of electric drives. The requirements of this methodology were:

- To take advantage of the information generated during the design phase to perform the fault analysis. During the design and validation phases, models of the system are used. These models contain all the information about the architecture and the behavior of the traction converter and should be used in the fault analysis process, as part of the model-based development of industrial products.
- Fault effects must be quantified. If the conclusions of the fault analysis are intended to condition the design, development and maintenance, these conclusions should be

quantitative. For example, in the field of electric drives, the effects should be described in terms of traction/braking capacity loss, consumption increase, harmonic component increase or comfort deterioration. There is also a lack of indicators for the detection and identification of faults.

- In order to achieve a better coordination between the design and safety teams, both should share the same tools. The main tool for the design and the safety analysis should be the model of the system. Nowadays the resources available in the development of traction systems are Matlab/Simulink for the simulation of models and HIL platforms for the validation of embedded traction control units.

In order to tackle these needs, in the next section, a new methodology for fault analysis is proposed. The methodology is based on models and an HIL platform, following the fault injection approach presented in Section 3.2.2. In the literature, most of the work presented about HIL platforms was focused on the development and validation of new control strategies and embedded units [27–29] and in some cases, authors mention the use of such platforms for the analysis of specific faults [30–33]. Nevertheless, it is difficult to find publications where an HIL platform is used systematically in the analysis of fault modes and effects of a system.

4.2. Hardware-in-the-loop platform

The methodology presented here uses an HIL platform to obtain information about the behavior of the faulty railway traction system and its control strategy. The structure of the system is shown in **Figure 5**.

The model of the traction system, designed in Matlab/Simulink, is simulated in an OPAL-RT Real-Time Simulator. Thanks to its analog and digital inputs and outputs, it communicates with a commercial TCU developed by CAF Power&Automation. The tests are monitored from an auxiliary PC where the data is stored and analyzed.

The Real-Time Simulator allows simulating the same models developed in the design stage. In the case of electric drives, these models are usually implemented using Matlab/Simulink, which is the language also used by the simulator. If the Real-Time Simulator requires the adaptation of the models, the implementation of the extended models itself becomes an objective, which is an obstacle in the integration of the design and the safety analysis tasks. In this case, as an OPAL-RT simulator is used, Matlab/Simulink models can be imported directly into this device, reducing the development and adaptation time, and enabling to reuse the existing know-how. **Figure 6** shows the simulation model for the traction system. This model is the same as the one used in the design stage and contains some additional blocks to manage the interaction between the simulator, the TCU and the monitoring PC. The extension of the model to make its execution in real time possible takes little time and there is no need for expert knowledge about real-time simulations.

Furthermore, it has to be taken into account that the aim of the simulations is to replicate the behavior of the system under faults, so a flexible simulation environment is needed. This environment has to allow an effective and simple way to simulate faults. The second

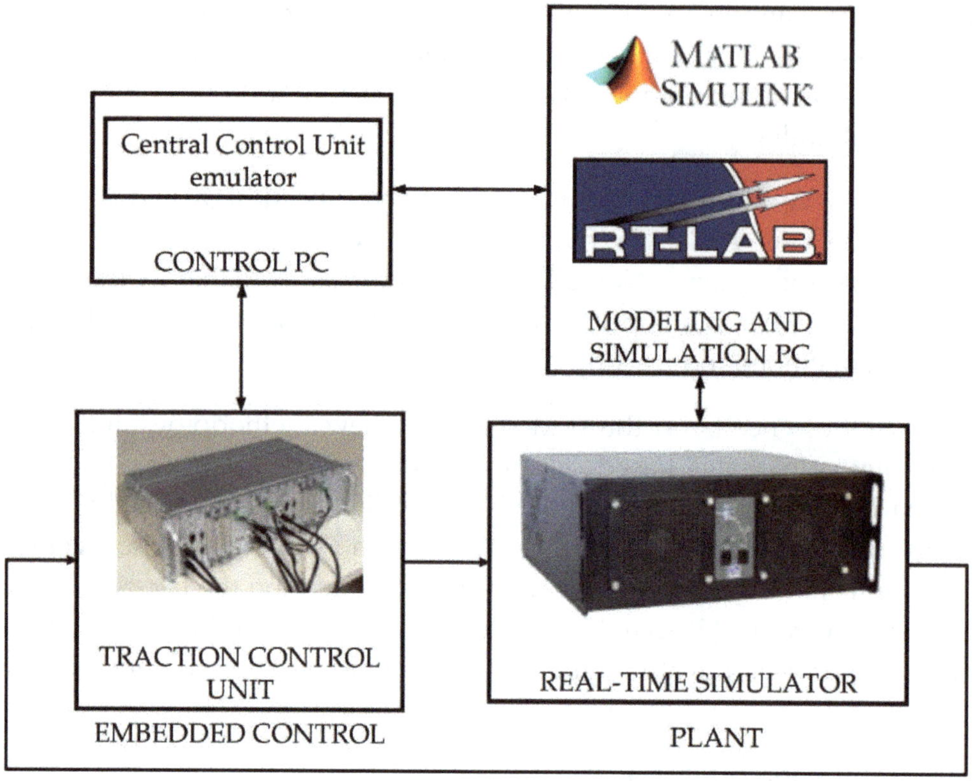

Figure 5. HIL platform structure.

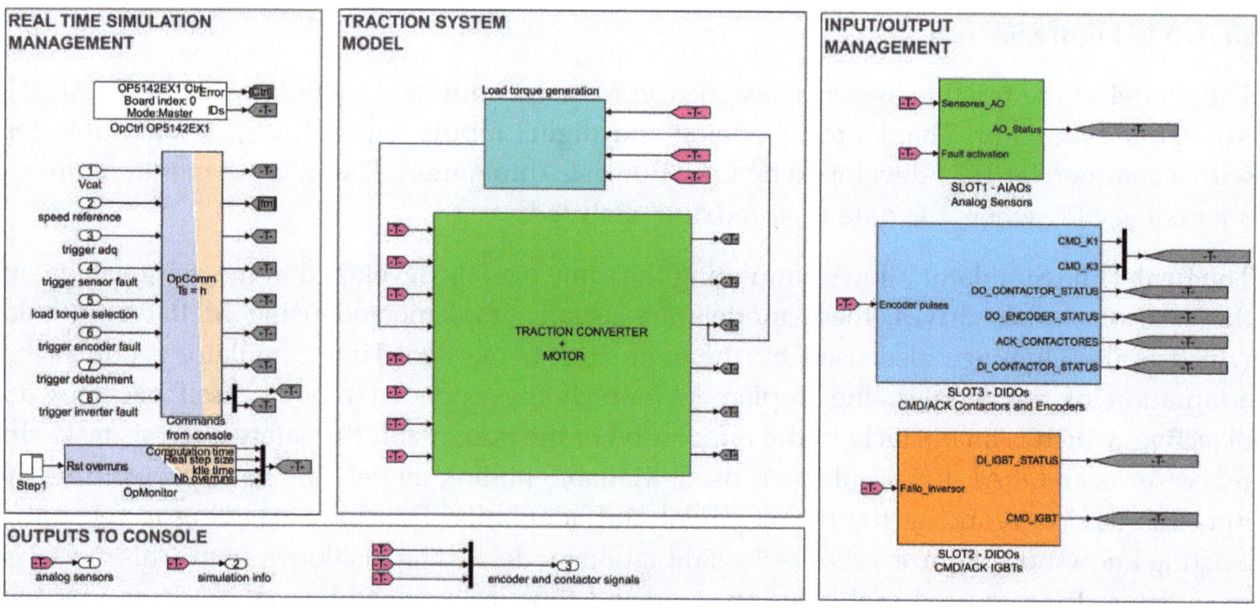

Figure 6. Real-time simulation model of railway traction system.

component of the HIL platform is the TCU. This device is the electronic control unit that controls the traction system. Thanks to the HIL platform, a commercial version of the TCU was used, executing the same control code as in a real application. This allows to gather information about the control strategy that otherwise, with a simple FMEA analysis, would be impossible to obtain. Moreover, using a commercial TCU, implementation and manufacturing details are taken into account and fed back into the enhanced FMEA. It is important to note that thanks to this kind of platforms the results of the fault analysis are quantitative and more detailed than the results obtained with classical methods.

4.3. Methodology

The starting point in the methodology for model-based fault analysis with HIL platforms is an FMEA based in conceptual/empirical knowledge about the system (see **Figure 7**). As conceptual FMEA we refer to the initial fault modes and effects analysis that is performed during the design stage of railway traction systems. This initial or existing FMEA is usually based on design and requirement documents and standards, following the classical methodologies for fault analysis.

The methodology is composed of the following steps:

1. Complete conceptual FMEA

   In the first step, all the information related to the architecture and the behavior of the traction system is gathered. The main information source is the initial FMEA (if there already was one), but other information sources should be considered:

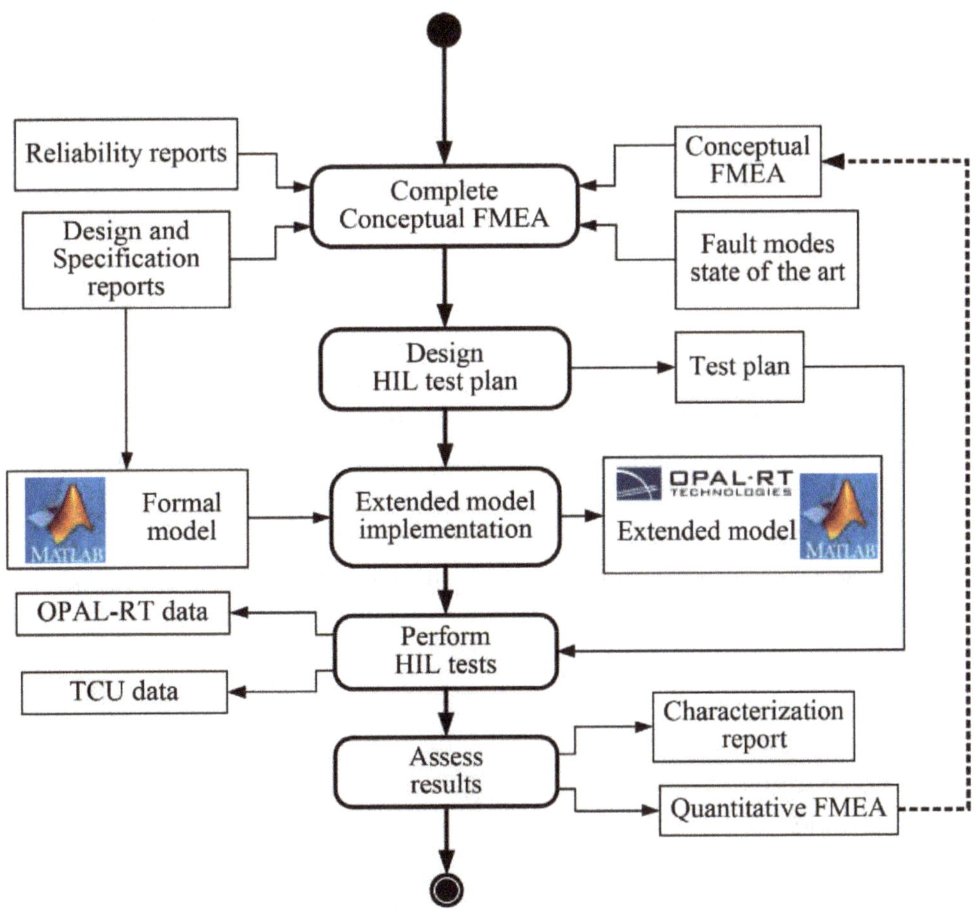

Figure 7. Flux diagram of the methodology for the model-based characterization of faults and effects.

- Design and specification documents where it is defined as what the system is (architecture) and how does it work (behavior).
- Reliability reports written by the after-sales department, where information about fault rates, mean time between failures and availability is presented.
- Updated state of the art about fault modes, causes and effects of the components of the traction system.

2. Design a test plan

   This plan is a document where all the conditions for each test are stated. Among others, these aspects should be defined:

   1. General description of the test
      a. Objective
      b. Fault mode: description and magnitude of the fault.
      c. Subsystem under fault. Part of the system that needs to be tested.
   2. Operation point. Operation phase where the system is working when the fault occurs (steady state, transient state, in traction, braking, and so on)
   3. Expected fault effects based on the conceptual FMEA. Thanks to the FMEA, preliminary fault effects are identified and the variables where the effects are visible established.
   4. Platform configuration. Define the configuration of the components of the HIL platform.
   5. Summary of the test to perform.

3. Implement and validate the extended model

   In this step, the formal model is extended in order to simulate the faulty behavior. This extended model is already defined by the fault modes that need to be tested and the expected effects considered in the test plan. The expected effects should be clearly stated in order to know if the extended model would be able to replicate them.

4. Perform HIL test

   HIL tests are performed following the test plan for the selected operation conditions, storing the required data.

5. Analyze and assess results

   A fault characterization report is written where the details of the faulty behavior are described. The changes that the fault has generated in the operation of the system are reported using tables and graphs. The data recorded in the TCU about the control strategy is analyzed and converted to quantitative information about the effects of the faults. Moreover, the conclusions are fed back to the initial FMEA, which is sent to the diagnostics and maintenance teams.

It is worth mentioning that this methodology can be applied iteratively adding information to the initial FMEA during the whole life cycle. It could be a tool for fault analysis throughout different phases of the life cycle. In this chapter, the model-based FMEA has been proposed as a tool to improve fault analysis during the validation process, in which a commercial TCU or a prototype is available for unit testing, as shown in **Figure 8**. In this diagram, a traction system development V-model is presented. The model describes the different phases of the life cycle. As it was mentioned, the HIL simulations are commonly used in the validation and testing process, but the methodology could be used in the design stage with model-in-the-loop (MIL) or software-in-the-loop (SIL) simulations.

With this methodology, fault analysis can be applied not only in the design phase but also in the system integration process, enhancing the quantity and the quality of the FMEA. It allows the continuous improvement of products in many aspects such as safety, maintenance and fault detection and identification functionalities.

An immediate result of the approach and the quantitative FMEA is the improvement of the maintenance manual, which is completed with quantitative information about the effects and the indicators of each fault.

4.4. Use case: quantitative FMEA of current sensors

In the following sections, a use case of the proposed methodology will be presented. In this case, the example will be focused on the analysis and identification of phase current sensor fault modes (FMs) and effects. The steps shown in **Figure 7** are explained for current sensor faults.

4.4.1. Complete conceptual FMEA

In this step, the initial FMEA is extended with additional information obtained from the literature and from the specification documents of the traction system (see **Table 2**).

First, it should be noted that the number of fault modes has increased. Some authors point out [34, 35] that an unbalanced measurement of three-phase currents in a traction drive generates a

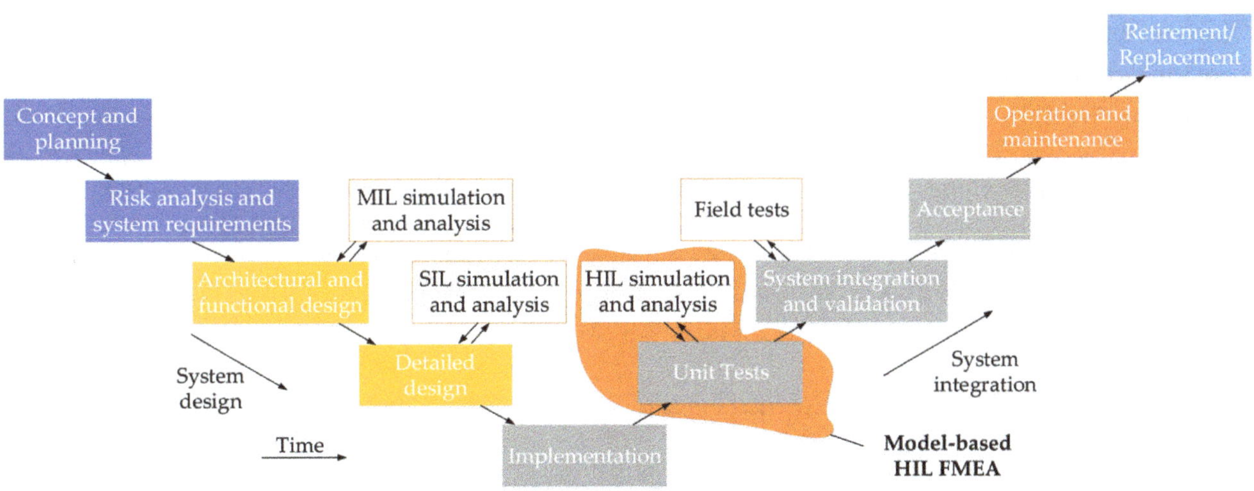

Figure 8. Railway traction system life cycle.

Operation phase	Fault mode	Cause	Local effect	Traction unit effect	Train effect
Normal operation	**FM1** Measured value bigger than real value (offset)	Internal failure: offset	False measurement	Oscillations in the torque at motor operation frequency	Loss of traction unit
Normal operation	**FM2** Measured value bigger than real value (gain)	Internal failure: gain	False measurement	Oscillations in the torque at twice stator frequency. Torque controlled below reference value	Loss of traction unit
Normal operation	**FM3** No measurement (sensor connected but no signal)	Internal failure	No measurement	Inappropriate control. Overcurrent. Disabled converter	Loss of traction unit
Start-up and motor fluxing	**FM4** Open-circuit	Internal failure	No measurement	Inappropriate fluxing. Overcurrent. Disabled converter	Loss of traction unit
Normal operation	**FM5** Open-circuit	Internal failure	False measurement	Inappropriate control. Overcurrent. Disabled converter	Loss of traction unit
Normal operation	**FM6** Measured value smaller than real value (offset)	Internal failure: offset	False measurement	Oscillations in the torque at stator frequency	Loss of traction unit
Normal operation	**FM7** Measured value smaller than real value (offset)	Internal failure: offset	False measurement	Oscillations in the torque at twice stator frequency. Torque controlled above reference value	Loss of traction unit

Table 2. Conceptual FMEA of a railway traction systems: Phase current sensor FMEA.

low-frequency oscillation in the torque. Depending on the type of deviation, offset or gain, the oscillation has different frequencies, and the effects change. Hence, the fault modes (FMs) have been described in detail.

Moreover, knowing that the control of the traction motor has different operation phases, it was considered interesting to analyze the effect of a sensor disconnection fault during the fluxing of the motor (FM4).

4.4.2. Design HIL test plan

Once the fault modes and effects are selected, the HIL test plan is defined. The test plan contains all the information mentioned in section 4.2 and will not be reproduced here due to lack of space.

4.4.3. Implement and validate the extended model

In this step, the extended model for the simulation of phase current sensor faults was implemented. Taking into account the fault modes described in **Table 2**, a fault injection bock was implemented to inject gain and offset faults in the measurement of the sensor (see **Figure 9**).

4.4.4. Perform HIL tests

The tests are performed following the test plan, for the required operation points and fault modes.

As a result, two data files about the evolution of the system are obtained. On the one hand, a file gathers the evolution of the internal variables of the traction system. In this case, the data is obtained from the real-time simulator. On the other hand, there is another group of variables that is stored and downloaded from the TCU. These variables reflect the behavior of the control strategy.

4.4.5. Assess results

In the evaluation step, the data from the tests is converted into information to improve and quantify the effects in the FMEA. In the case of the phase current sensor faults, two effects were identified. Due to gain and offset deviations, a new harmonic component appears in the

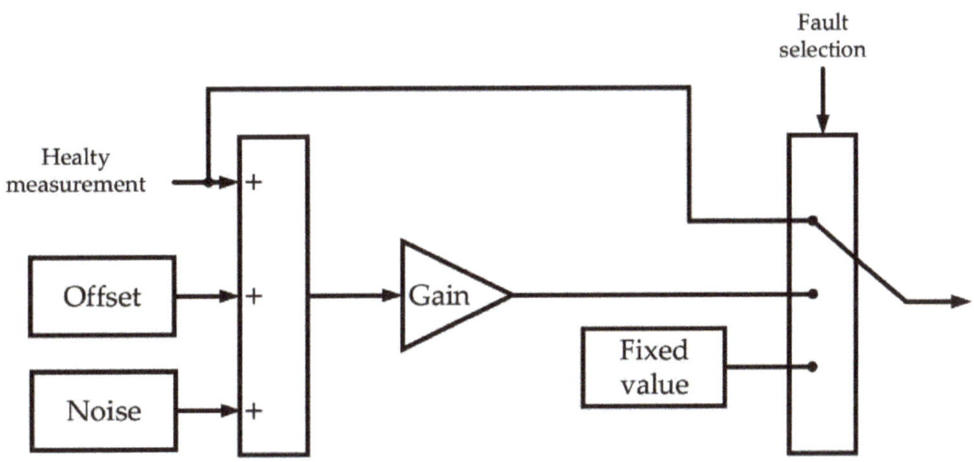

Figure 9. Fault injection block for phase current sensors.

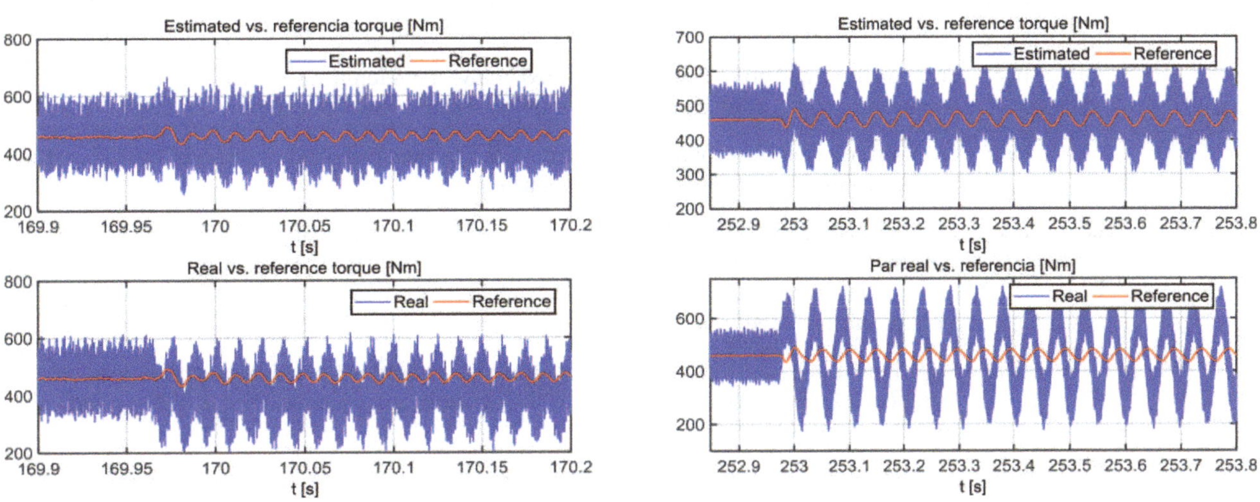

Figure 10. Reference, real and estimated torque with phase current sensors under +20% gain deviation (left) under 100A offset deviation (right).

torque, as it is shown in **Figure 10**. If the deviation is caused by an offset, the oscillation has the same frequency as the supply current. This frequency is twice the supply frequency when the fault mode is a gain deviation. Moreover, when the gain deviates, the torque is controlled below or above the reference value.

With this methodology, each fault mode has been characterized by different deviation levels. For example, as it is presented in **Figure 11** the relation between de gain deviation and the torque oscillations and deviations was obtained.

Table 3 shows the summary of the results.

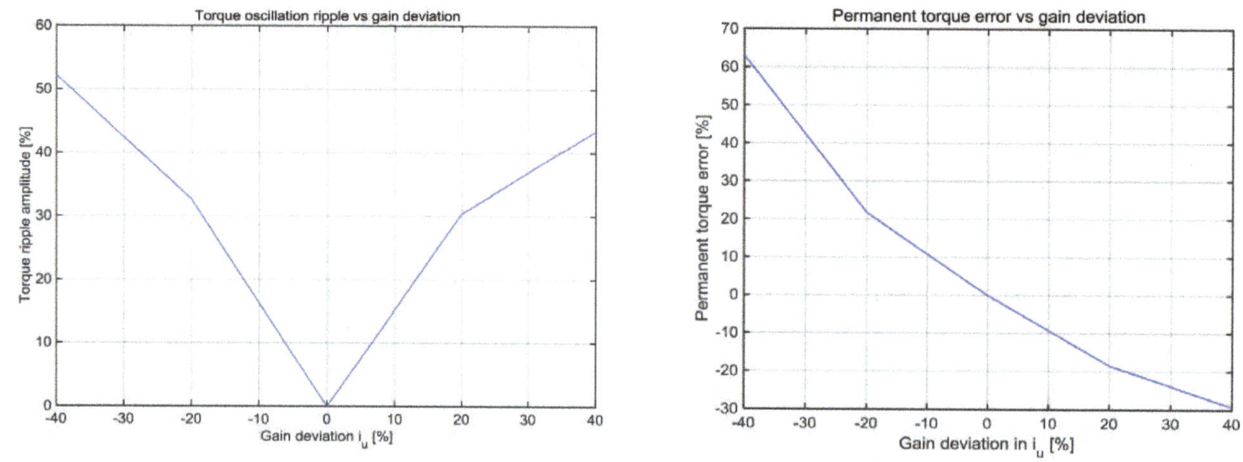

Figure 11. Phase sensor gain deviation due to phase sensor fault (left) torque ripple(right) torque permanent deviation.

Traction unit level effect	Effect 1	Oscillation in the torque and the bus voltage at twice the supply frequency of the motor. The quantitative amplitude of the torque ripple vs. the gain deviation is shown in **Figure 11**.
	Effect 2	Permanent torque error. Torque controlled below the reference value. The quantitative amplitude of the torque ripple vs. the gain deviation included in **Figure 11**.
	Effect 3	Overcurrents and overvoltages that activate the protections. Above ±35% of deviation, the inverter is disabled to assure component safety.
Train level effect	Effect 1	Comfort loss at low speeds. Low-frequency oscillations in the torque. Using the torque ripple curves, for each railway application, a quantitative evaluation of oscillations of the train linear acceleration could be computed by obtaining the maximum allowable torque ripple for a good comfort travel.
	Effect 2	Maximum acceleration capacity is decreased. Torque capacity lost. For each application, using the information of the train route, the overtime duration associated with a gain fault could be assessed.
	Effect 3	Availability loss of the one Traction Converter above ±35% of deviation of the gain fault.

Table 3. Quantitative FMEA for phase current sensor under gain faults.

5. Conclusions

In this chapter, a methodology for model-based HIL fault analysis was presented. Using models and real-time simulations, an improved quantitative FMEA for complex systems can be obtained. Following the described steps, a quantitative FMEA for railway traction systems was obtained as an example. Thanks to the data obtained from the models and the simulations, the effects of the faults are characterized in detail. The improved FMEA can be used as a reference document to improve designs, to implement new diagnostic functionalities or to elaborate new maintenance procedures.

Acknowledgements

This research work was supported by CAF Power & Automation. The authors are thankful to the colleagues from CAF Power & Automation, who provided expertise that greatly assisted the research.

Author details

Jon del Olmo*, Fernando Garramiola, Javier Poza and Gaizka Almandoz

*Address all correspondence to: jdelolmo@mondragon.edu

Electronics and Computing Department, Mondragon Unibertsitatea, Faculty of Engineering, Arrasate-Mondragón, Spain

References

[1] Jardine AKS, Lin D, Banjevic D. A review on machinery diagnostics and prognostics implementing condition-based maintenance. Mechanical Systems and Signal Processing. 2006;**20**(7):1483-1510

[2] Farnsworth M, Tomiyama T. Capturing, classification and concept generation for automated maintenance tasks. CIRP Annals - Manufacturing Technology. 2014;**63**(1): 149-152

[3] Le Mortellec A. Proposition d'une architecture de surveillance active à base d'agents intelligents pour l'aide à la maintenance de systèmes mobiles-Application au domaine ferroviaire. Valenciennes: Université de Valenciennes et du Hainaut-Cambrésis; 2014

[4] Gandibleux J. Contribution a l'evaluation de surete de fonctionnement des architectures de surveillance diagnostic embarquees Application au transport ferroviaire. Valenciennes: Université de Valenciennes et du Hainaut-Cambresis; 2013

[5] Xue F, Yan W, Roddy N, Varma A. Operational data based anomaly detection for locomotive diagnostics. In: International Conference on Machine Learning, Models, Technologies and Applications. MLMTA. 2006. pp. 236-241

[6] European Commission, "Shift2Rail Joint Undertaking Multi-Annual Action Plan," Shift2Rail Joint Undertaking. Brussels. 2015

[7] Kabir S. An overview of fault tree analysis and its application in model based dependability analysis. Expert Systems with Applications. 2017;**77**:114-135

[8] Sharvia S, Kabir S, Walker M, Papadopoulos Y. Chapter 12 - Model-based dependability analysis: State-of-the-art, challenges, and future outlook. In: Software Quality Assurance. Boston: Morgan Kaufmann. 2016. pages 251-278

[9] Joshi A, Heimdahl MPE. Model-based safety analysis of simulink models using SCADE design verifier. In: SAFECOMP 2005 – International Conference on Computer Safety, Reliability and Security. LNCS. Vol. 3688. Berlin, Heidelberg: Springer. 2005. pp. 122-135

[10] Lisagor O, Pumfrey DJ, Mcdermid JA. Towards a practicable process for automated safety analysis. In: 24th International System Safety Conference (ISSC) Organized by The International System Safety Society. 2006. pp. 596-607

[11] Villacourt M. Failure mode and effects analysis (FMEA): A guide for continuous improvement for the semiconductor equipment industry. SEMATECH. 1992

[12] Liu H, Liu L, Liu N. Expert systems with applications risk evaluation approaches in failure mode and effects analysis : A literature review. Expert Systems with Applications. 2013;**40**(2):828-838

[13] Mikulak RJ, McDermott R, Beauregard M. The Basics of FMEA. 2nd ed. New York: CRC Press; 2008

[14] David P, Idasiak V, Kratz F. Reliability study of complex physical systems using SysML. Reliability Engineering & System Safety. 2010;**95**(4):431-450

[15] Sharvia S, Papadopoulos Y. Integrated application of compositional and behavioural safety analysis. In: Dependable Computer Systems. Vol. 97. Berlin, Heidelberg: Springer; 2011. pp. 179-192

[16] Belmonte F, Soubiran E. A model based approach for safety analysis. In: SAFECOMP 2012 – International Conference on Computer Safety, Reliability and Security. LNCS. Vol. 7613. Berlin: Springer; 2012. pp. 50-63

[17] Grunske L, Winter K, Ytapanage N, Zafar S, Lindsay PA. Experience with fault injection experiments for FMEA. Software – Practice and Experience. 2011;**41**:1231-1258

[18] Aizpurua JI, Muxika E. Model-based design of dependable systems: Limitations and evolution of analysis and verification approaches. International Journal on Advances in Security. 2013;**6**(1):12-31

[19] Sharvia S, Papadopoulos Y. Integrating model checking with HiP-HOPS in model-based safety analysis. Reliability Engineering & System Safety. 2015;**135**:64-80

[20] Lisagor O. Failure Logic Modelling: A Pragmatic Approach. York: University of York; 2010

[21] Papadopoulos Y, Maruhn M. Model-based synthesis of fault trees from Matlab-Simulink models. In: 2001 International Conference on Dependable Systems and Networks. Vol. 36. 2001. pp. 77-82

[22] Papadopoulos Y. Safety-Directed System Monitoring Using Safety Cases. York: The University of York; 2000

[23] Kabir S. Compositional Dependability Analysis of Dynamic Systems with Uncertainty. Hull: The University of Hull; 2016

[24] Bozzano M et al. ESACS: An integrated methodology for design and safety analysis of complex systems. In: Proceedings of the European Safety and Reliability Conference 2003, ESREL2003. 2003

[25] Akerlund O, Bieber P, Böde E. ISAAC, a framework for integrated safety analysis of functional, geometrical and human aspects. In: 3rd European Congress on Embedded Real Time System (ERTS). 2006

[26] Estefan JA. Survey of model-based systems engineering (MBSE) methodologies. In: International Council on Systems Engineering (INCOSE). 2008

[27] Terwiesch P, Keller T, Scheiben E. Rail vehicle control system integration testing using digital hardware-in-the-loop simulation. IEEE Transactions on Control Systems Technology. 1999;**7**(3):352-362

[28] Wang L, Zhang Y, Yin C, Zhang H, Wang C. Hardware-in-the-loop simulation for the design and verification of the control system of a series-parallel hybrid electric city-bus. Simulation Modelling Practice and Theory. 2012;**25**:148-162

[29] Isermann R, Schaffnit J, Sinsel S. Hardware-in-the-loop simulation for the design and testing of engine-control systems. Control Engineering Practice. 1999;**7**(5):643-653

[30] Poon JJ, Kinsy MA, Pallo NA, Devadas S, Celanovic IL. Hardware-in-the-loop testing for electric vehicle drive applications. In: Conference Proceedings – IEEE Applied Power Electronics Conference and Exposition – APEC. 2012. pp. 2576-2582

[31] Wu J, Dufour C, Sun L. Hardware-in-the-loop testing of hybrid vehicle motor drives at Ford Motor Company. In: 2010 IEEE Vehicle Power and Propulsion Conference (VPPC 2010); 2010

[32] Baccari S et al. Real-time hardware-in-the-loop in railway: Simulations for testing control software of electromechanical train components. In: Railway Safety, Reliability and Security: Technologies and Systems. 2012. p. 487

[33] Alvarez-gonzalez F, Member S, Griffo A, Wang J, Member S. Real-time hardware-in-the-loop simulation of permanent magnet synchronous motor drives under stator faults. IEEE Transactions on Industrial Electronics. 2017;**64**(9):6960-6969

[34] Chung D-WCD-W, Sul S-KSS-K. Analysis and compensation of current measurement error in vector-controlled AC motor drives. IEEE Transactions on Industry Applications. 1998;**34**(2):340-345

[35] Jung H, Kim J, Kim C, Choi C. Diminution of current measurement error for vector controlled AC motor drives. IEEE Transactions on Industry Applications. 2006;**42**(5): 1249-1256

8

Vibration Mitigation of Railway Bridge Using Magnetorheological Damper

Muzaffer Metin, Arif Ulu, Mahmut Paksoy and Murat Emre Yücel

Abstract

The purpose of this study is to analyze the railway bridge vibrations and control their negative effects through semi-active magnetorheological (MR) damper. Dynamic analysis of a railway bridge subjected to the moving load is performed. The real structural parameters are used, and the six-axle train is simulated as moving loads. The railway bridge is modeled as Euler-Bernoulli beam theory, and it is discretized through Galerkin method. To mitigate the bridge vibrations, MR damper with a fuzzy logic-based controller (FLC) is positioned at the ends of the bridge. The simulations of the system are performed by MatLab software. Finally, the results are examined both in the time and frequency domains.

Keywords: bridge vibration, vibration control, semi active control, magnetorheological damper, adaptive control

1. Introduction

In general, because of the increasing air pollution and traffic problems, rail vehicles gained importance as a mass transportation system. While the transportation speeds increased with the development of technology, expectations of comfort are also raised along the entire line, including bridges and viaducts. On the other hand, the bridges are enabled to be built light and slender. This made the bridges prone to the vibrations triggered by moving rail vehicles. The resulting vibrations reduce the safety of travel while also affecting the comfort of the passengers negatively. For this reason, the vibration analyses of railway bridges are considered as a significant factor in bridge design [1]. In recent years, research and development activities on suppressing railway bridge vibrations are increasingly concentrated, especially to increase passenger comfort without compromising safety at high speeds [1, 3, 11, 12]. For conventional speeds in railway transport, it may be sufficient to apply only passive methods such as polyurethane materials to insulate the bridge vibrations. However, after the increasing speeds on the railways, semi-active and active control methods are begun to be tried as only passive methods do not provide the desired performance in suppressing bridge vibrations [1, 3, 11, 12].

The dynamic behavior of railway bridges under moving load is a complicated and challenging phenomenon and has drawn the attention of scientists and engineers due to the complex structure of railway bridges. The interaction between rail vehicle and bridge creates a dynamic effect. The most crucial parameters determining the dynamic response of bridges are rail vehicle speed, characteristics of bridge, and rail irregularity. The comfort level of the rail vehicles must meet expectations, while the safety of running is at the highest level. In order to do so, suppressing the railway bridge vibrations is significant as well as rail vehicle suspensions. In addition, the vibration control of a rail bridge is better both for bridge's life and safety of the rail vehicles. This also brings passengers' comfort improvement and allows them to pass faster.

Structural damping is one of the typical characteristics that damps the vibration effect of structures. Yet, that damping is regarded as insufficient. So, when the disturbance force is applied, it may cause strong and long-lasting vibrations. Hence, passive, semi-active, and active suspensions to mitigate vibrations are investigated.

Related literature shows that the structural control of the railway bridges subjected to the moving load is studied by many researchers. The bridge can be modeled as a simply supported Euler-Bernoulli beam [2], and the train mass is modeled at a constant speed as a time and spatially changing load. For this aim, lumped parameters of the vehicle can be neglected [3]. Also, some models which have lumped parameters are adopted subsequently in the related study [4, 5]. On the other hand, several researches on bridges' dynamic response under the moving load demonstrate the effects of moving train.

The suspension types of railway bridges are separated into three groups which are passive, semi-active, and active suspensions. However, only the active suspensions can be controlled by applying an external force. The idea in implementing a semi-active suspension is to change active force generator with adaptive elements that can shift the rate of energy dissipation in response to a momentary condition of motion. The force of suspension can be controlled through active causes in response to sensory feedback, whereas the actuators are used in active controllers to implement an independent force on suspension [6]. We could say that semi-active systems are more practical than passive systems and less expensive and complicated than active systems [7]. It is widely known that the MR damper is quite feasible and reliable to implement in reducing vibrations [8] since its performance is better than passive suspension as its power requirements are low and its hardware is less expensive than active suspension [9]. Usually, the MR damper-based semi-active controller works through a two-step progress. Firstly, a system controller designates the desired control force in respect of the responses; then damper controller sets the command applied to the MR damper so that it can track the desired control force. Hence, the success of MR damper-based semi-active controller depends on two aspects: One of them is to select a proper control strategy, and the other is to establish the accurate damper controller [10].

In this paper, the vibration of railway bridges subjected to the moving load is investigated. The bridge model is taken into consideration as a simple support beam. As the model is a continuous one, it is changed to discrete model through the Galerkin method, and its vibration is investigated when subjected to the force which is due to the train passing on the bridge. To mitigate vibrations, two symmetric MR dampers are applied to the bridge from the bottom. Fuzzy logic control method is used on MR dampers to determine the voltage input. Controlled and uncontrolled results of vibration analyses are analyzed.

2. Mathematical modeling

Figure 1 shows a model of railway bridge. Bridge modeled as Euler-Bernoulli beam is a constant cross section, homogeneous, and simply supported. At that time MR dampers that modeled as modified Bouc-Wen model is located on two sides of bridge. In addition, forces are thought as axial forces of railway train axles:

$$m\frac{\partial^2 w(x,t)}{\partial t^2} + c\frac{\partial w(x,t)}{\partial t} + EI\frac{\partial^4 w(x,t)}{\partial x^4} = \sum_{j=1}^{6}\delta(x-vt)P_j + \sum_{j=1}^{2}\delta(x = x_{dj})F_{MRj} \tag{1}$$

where EI, m, c, and $w(x, t)$ were flexural rigidity mass per length, the damping coefficient, and transverse displacement of bridge at point x and time t, respectively. Parameters of bridge were given in **Table 1**. Right hand of equation is axial forces (P) represented by Dirac-delta function

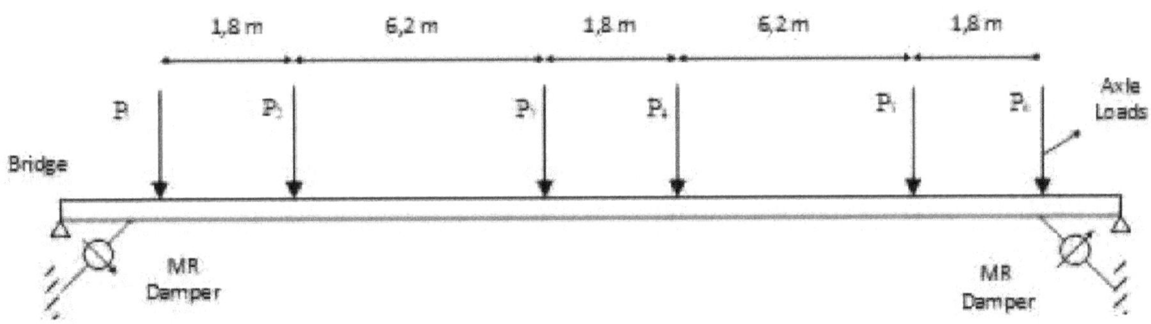

Figure 1. Railway bridge with MR dampers acted on moving axle loads.

Young's modulus (N/m^2)	210×10^9
Area moment of inertia (m^4)	0.61
Mass per length (kg/m)	18,400
Length of the beam (m)	42
Moving load (N)	80,000
Damping ratio	0.1
MR damper locations (m)	5.37

Table 1. Properties of railway bridge [11].

and MR damper forces. Dirac-delta function $\delta(x)$ was thought as a unit concentered force acting at point $x = 0$. Dirac-delta function was defined in Eq. (2):

$$\int_a^b \delta(x - vt)f(x)dx = f(\xi) \quad \text{for} \quad a < \xi < b \tag{2}$$

By using Galerkin method, transverse function $w\ (x,\ t)$ was transformed into two separate functions Eq. (3). A sinus function $\left(\sin\frac{i\pi x}{L}\right)$ depending on x, in order to satisfy boundary conditions, was selected:

$$w(x,t) = \sum_{i=1}^{N} T(t) \sin\frac{i\pi x}{L} \qquad i = 1, 2, \ldots, N \tag{3}$$

Trial function (3) is implemented in Eq. (1), multiplied full equation with trial function and integrated from *0* to *L*. Finally, a partial differential equation (Eq. (1)) is turned into ordinary differential equation (Eq. (4)):

$$m\ddot{T}_i(t) + c\dot{T}_i(t) + EI\left(\tfrac{i\pi}{L}\right)^4 T_i(t) = \sum_{j=1}^{6} \frac{2P_j}{L} \sin\left(\frac{i\pi(vt - L_j)}{L}\right) + \sum_{j=1}^{2} \frac{2F_{MRj}}{L} \sin\left(\frac{i\pi x_{dj}}{L}\right) \quad i = 1, 2, 3, \ldots, N \tag{4}$$

At the right-hand side of equation, there are eight forces. The first six forces are moving forces that represented train axle loads. L_1, L_2… are distance from first wheel (**Figure 1**). The last ones are the MR damper forces that were located two sides (x_{d1}, x_{d2}) of bridge.

Damping of bridge is modeled as Rayleigh structural damping [2] and depends on mass, rigidity, and natural frequencies of the bridge (Eqs. (5)–(6)). ω_i and ω_j are represented natural frequencies of the simply supported bridge. Railway bridge model parameters are given in **Table 1**:

$$c = a_0 m + a_1 k \tag{5}$$

$$a_0 = \xi \frac{2\omega_i \omega_j}{\omega_i + \omega_j};\ a_1 = \xi \frac{2}{\omega_i + \omega_j} \tag{6}$$

As mentioned above, MR Dampers modeled modified Bouc-Wen Model that related equations are given Eqs. (7)–(12) [12]:

$$\dot{z} = -\gamma|\dot{x} - \dot{y}||z|^{n-1}z - \beta(\dot{x} - \dot{y})|z|^n + A(\dot{x} - \dot{y}) \tag{7}$$

$$\dot{y} = \frac{1}{c_0 + c_1}[az + c_0\dot{x} + k_0(x - y)] \tag{8}$$

A (m^{-1})	2769
β, γ (m^{-1})	647.46
k_0 (N/m)	137,810
n	10
x_0 (m)	0.18
k_1 (N/m)	617.31

Table 2. Model parameters of MR damper [12].

where k_1 is the accumulator stiffness, c_0 is the viscous damping at larger velocities, c_1 is viscous damping for force roll-off at low velocities, x_0 is the initial displacement of spring k_1, and A, β, γ, and n are the constants about MR damper. The force was calculated as Eq. (9) [12]. Model parameters of MR damper is given in **Table 2**:

$$\begin{aligned} F_{MR} &= az + c_0(\dot{x} - \dot{y}) + k_0(x - y) + k_1(x - x_0) \\ &= c_1\dot{y} + k_1(x - x_0) \end{aligned} \tag{9}$$

c_0, c_1, and a have form of third-order polynomial with respect to electrical current i, expressed as Eqs. (10)–(12):

$$a(i) = 16566i^3 - 87071i^2 + 168326i + 15114 \tag{10}$$

$$c_0(i) = 437097i^3 - 1545407i^2 + 1641376i + 457741 \tag{11}$$

$$c_1(i) = -9363108i^3 + 5334183i^2 + 48788640i - 2791630 \tag{12}$$

Axle loads of Asea Brown Boveri (ABB) brand light rail vehicle used in the Istanbul urban transportation are considered as moving loads (**Figure 2**). Railway vehicle parameters are given in **Table 3**.

Figure 2. ABB railway vehicle.

Length (m)	23.2
Width (m)	2.65
Passengers capacity	257
Max design axle load (kN)	80
Wheel diameter (m)	0.68–0.6
Wheel width (m)	0.125
Max speed (km/h)	80

Table 3. ABB railway vehicle parameters [13].

3. Fuzzy control design

Fuzzy logic-based controllers (FLC) are frequently used in vibration reduction problems. Classical fuzzy logic controller is used in this paper which is based on two-input one-output FLC structure. The overall structure of used controller is shown in **Figure 3**.

The structure of fuzzy logic controller has two inputs and one output. The inputs are, respectively, "V_1" which is defined as the velocity of middle point of bridge model, and "V_2" which is defined as the velocity of the upper end point of MR damper. Linguistic variables which imply inputs and output are classified as NB NM NS ZO PS PM PB. Inputs and output are all normalized in the interval of [−1, 1], as well as outputs are normalized at range of [0, 1] as shown in **Figure 4**. Linguistic values which are used as output values are the following: ZO, VS, S, SM, M, B, and VB.

The variables are scaled with coefficient of S_{V1}, S_{V2}, and S_u. The fuzzy control rule is in the form of:

IF e = E_i **and** de = dE_j **than** V = $V_{(i,j)}$.

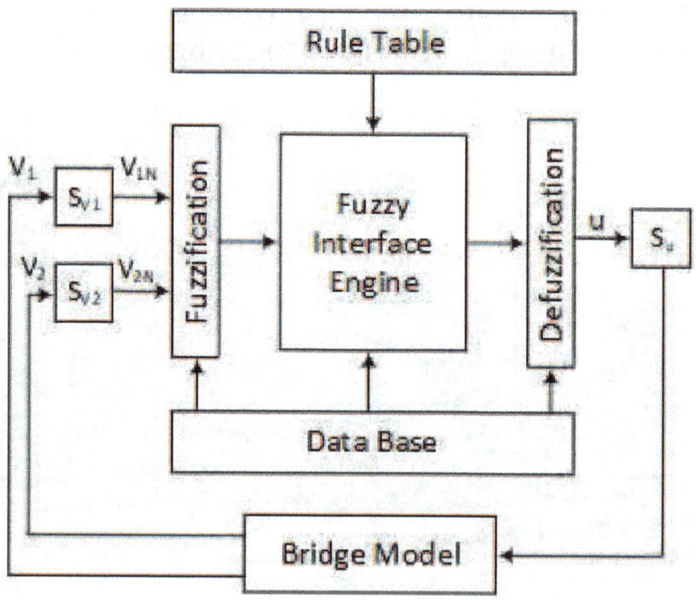

Figure 3. Block diagram of the two-input one-output fuzzy logic controller.

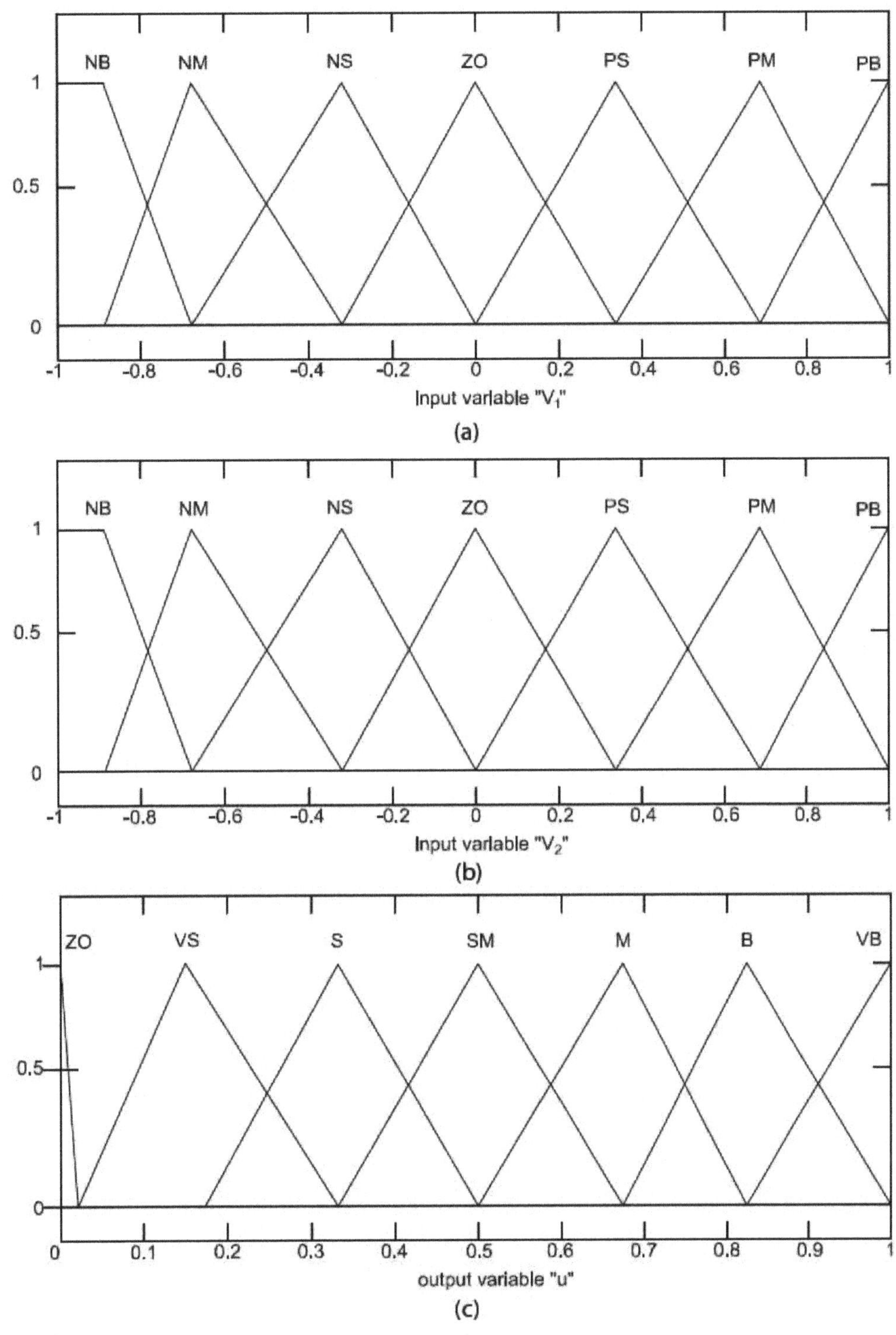

Figure 4. Membership functions of inputs V_1 (a) and V_2 (b) and output u (c).

These rules are written in a rule base lookup table which is shown in **Table 4**. The rule base structure is Mamdani type.

The linguistic labels used to describe the Fuzzy sets are "negative big" (NB), "negative medium" (NM), "negative small" (NS), "zero" (ZO), "positive small" (PS), "positive medium" (PM), "positive big" (PB), very small (VS), small (S), small medium (SM), medium (M), big (B), and very big (VB). It is possible to assign the set of decision rules as shown in **Table 1**. These

V1/V2	NB	NM	NS	ZO	PS	PM	PB
NB	SM	S	VS	ZO	VS	S	SM
NM	M	SM	VS	ZO	VS	SM	M
NS	B	M	S	ZO	S	M	B
ZO	VB	B	SM	ZO	SM	B	VB
PS	B	M	S	ZO	S	M	B
PM	M	SM	VS	ZO	VS	SM	M
PB	SM	S	VS	ZO	VS	S	SM

Table 4. Rules of fuzzy logic controller.

rules contain the input-output relationships that define the control strategy. Each control input has 7 fuzzy sets, so that there are 49 fuzzy rules.

4. Simulations

According to Eq. (4), the bridge model is considered for five modes. It was enough for the reliable responses of the bridge. MR damper models and bridge equations with fuzzy control tools are simulated in MATLAB-Simulink and performed in ode45 solver.

In all analysis train's speed was fixed to a maximum of 80 km/h for urban transportation. Moving loads are acting on bridge during 2.853 s. In uncontrolled system, supplied electrical current is fixed 0.05 A.

Figures 5–7 show the dynamic responses of midpoint of the railway bridge. Blue straight and red dashed lines show the uncontrolled and fuzzy controlled system, respectively. Maximum values of bridge responses are suppressed successfully, especially in velocity and acceleration. Also, settlement time of controlled bridge vibrations turns out to be better than uncontrolled system:

Figures 8–9 show the dynamic MR damper forces and electrical current that supplied the dampers. Straight lines and dashed lines represent the responses of the first and second MR dampers. **Figures 5**, **6**, and **8** show that the MR dampers generate forces in the same direction with the bridge velocity and in the opposite direction with the bridge motion. To control the MR damper structure, the maximum current occurs at 1.8 A.

If it is considered that displacement is concerned with running safety and acceleration with passenger comfort, **Figure 5** and **7** show that the MR damper performance is quite good in terms of both safety and comfort.

Power spectral density (PSD) of bridge vertical acceleration is shown in **Figure 10**. When we analyze **Figure 10**, it can be seen that the fuzzy logic controller reduced the magnitude of the bridge vertical acceleration in all frequencies significantly.

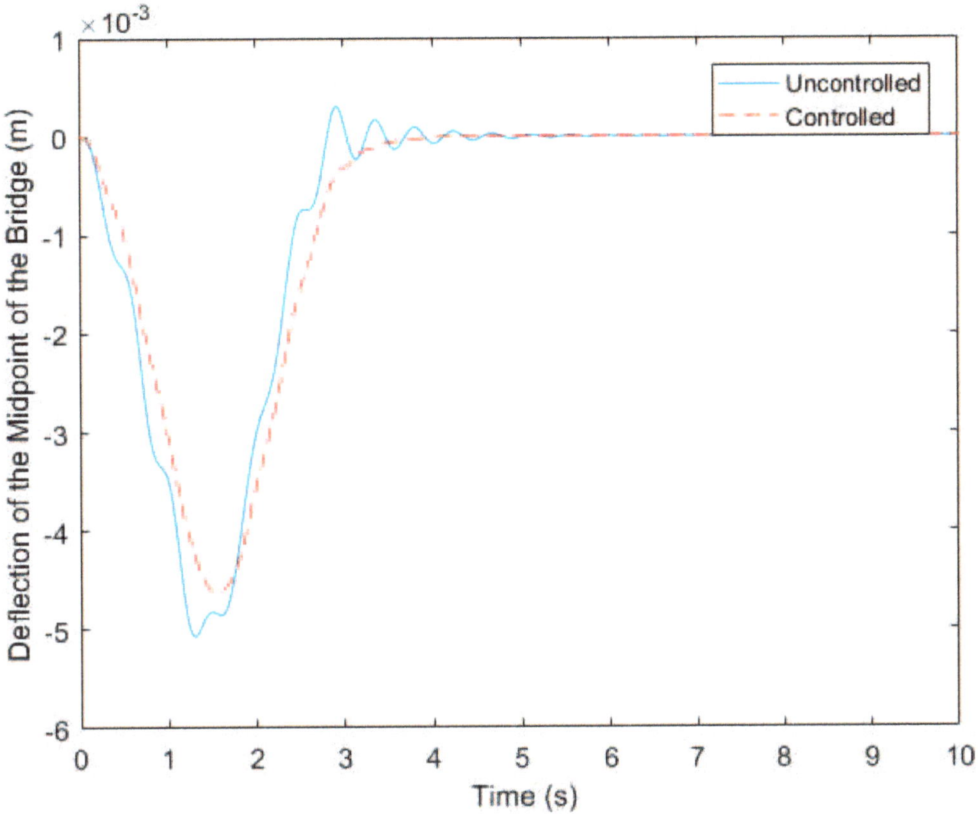

Figure 5. Displacement of the midpoint of the railway bridge.

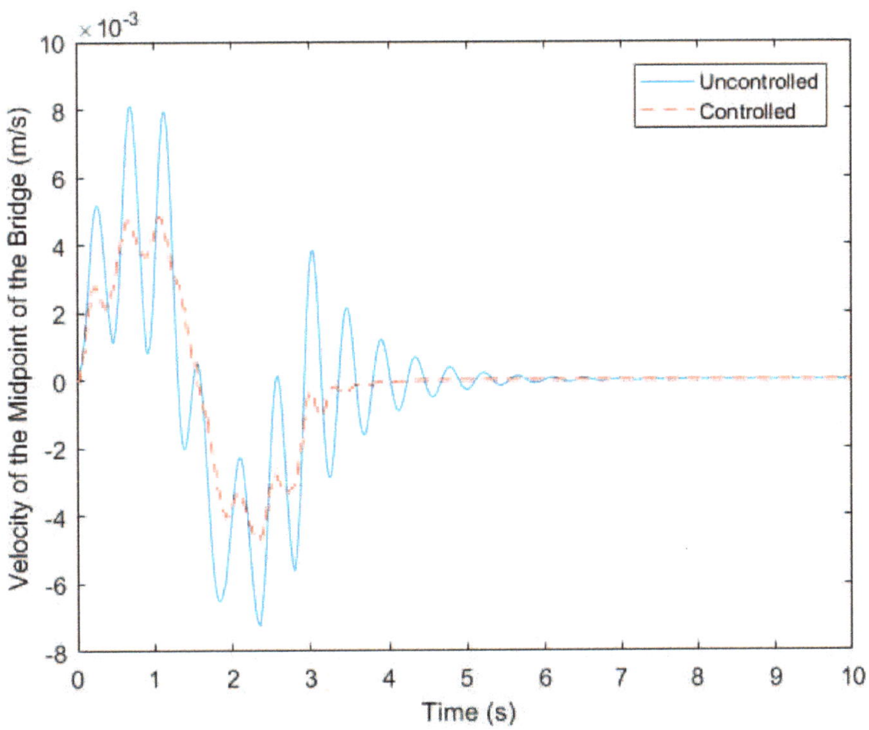

Figure 6. Velocity of the midpoint of the railway bridge.

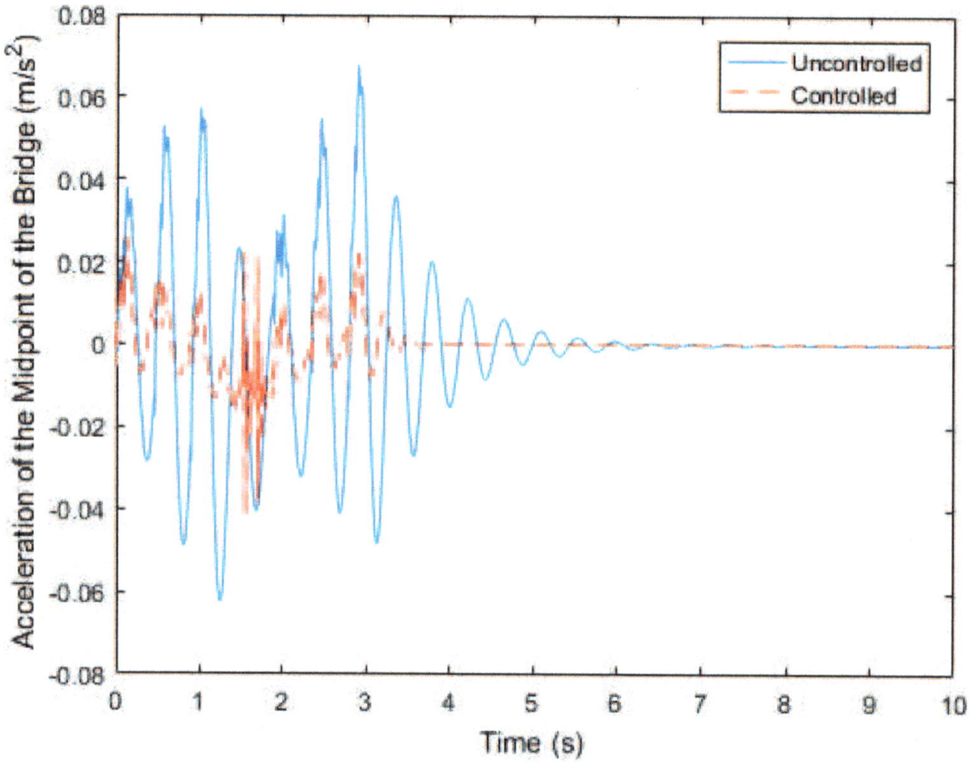

Figure 7. Acceleration of the midpoint of the railway bridge.

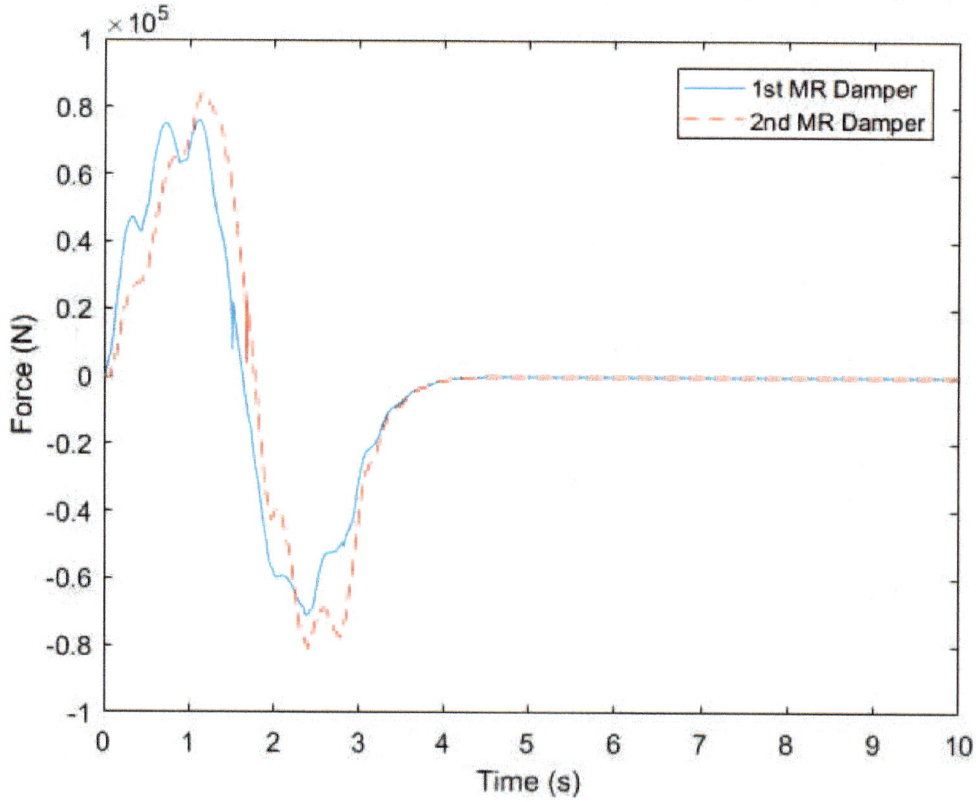

Figure 8. Two MR dampers' forces.

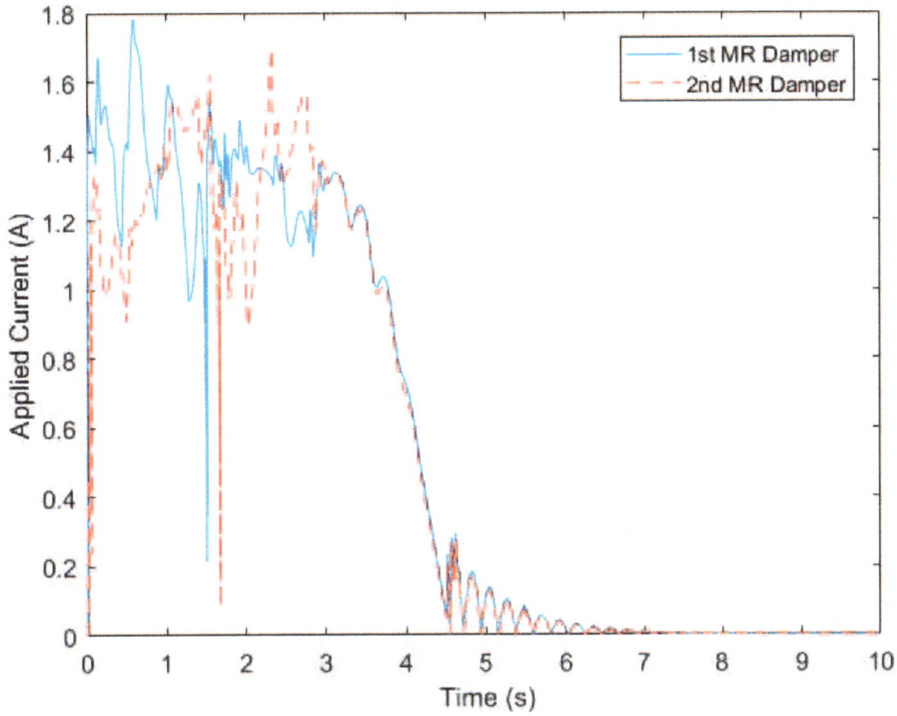

Figure 9. Applied current on the MR dampers.

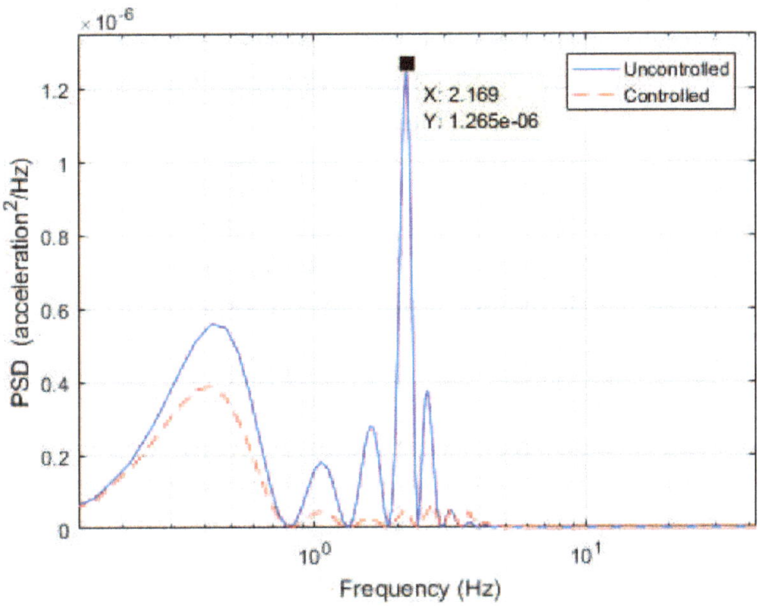

Figure 10. Power spectral density of acceleration data of the railway bridge.

Natural frequencies are calculated as 2.29, 9.17, 20.63, 36.69, and 57.32 Hz, by using Eq. (13). It is widely known that the most dangerous frequency is the first natural frequency in the structures. In this regard, the first natural frequency is well suppressed via fuzzy logic-controlled MR dampers:

$$W_n = \left(\frac{i\pi}{L}\right)^2 \sqrt{\frac{EI}{m}}\ i = 1, ..5 \tag{13}$$

5. Conclusions

The responses of a passive system to track inputs or other disturbances are obtained by using parameters such as inertia, spring, and damping. However the response for an active suspension system is developed by using a control algorithm. In this study the railway bridge vibrations are controlled through magnetorheological (MR) damper by using fuzzy logic control algorithms. This controller is preferred because of its superior performance in semi-active vibration control.

Firstly, the railway bridge is modeled as Euler-Bernoulli beam. Equations of the bridge are achieved according to Galerkin method. To suppress the railway bridge vertical vibrations, two MR dampers are positioned at the bridge ends. In the mathematical model of the bridge, MR damper is considered as friction-based modified Bouc-Wen model. The damping force of MR damper which changes with applied electrical current is controlled by the use of fuzzy logic controller. This control method's high performance, easy design, and robust character are some of the reasons for using it.

In simulations, railway vertical vibrations are analyzed for active and passive MR damper situations while six-axle railway vehicle is passing through the bridge. When the simulation results are examined, the vibration reduction performance of fuzzy logic controller in time and frequency domain can be seen. FLC performance has been simulated by comparing the results of passive and semi-active MR damper models.

Extension of bridge life, mitigation of negative effect of vibrations on human bodies and rail vehicles, and increasing of passenger comfort can be provided by reduction of bridge displacements and accelerations when passing the railway vehicle. It is observed that the method used in this study shows superior performance in the simulation environment and produces results that are suitable for all these purposes.

For the future work, fuzzy logic controller performance should be investigated taking into account the rail roughness resulting from wearing. In this way, adaptive methods for input and output parameters of controller should be developed to improve the controller performance. Also, different control algorithms can be compared, and it can be applied on the real bridge systems.

Author details

Muzaffer Metin*, Arif Ulu, Mahmut Paksoy and Murat Emre Yücel

*Address all correspondence to: mmetin@yildiz.edu.tr

Department of Mechanical Engineering, Yildiz Technical University, Turkey

References

[1] Luu M, Martinez-Rodrigo MD, Zabel V, Könke C. Semi-active magnetorheological dampers for reducing response of high-speed railway bridges. Control Engineering Practice. 2014; **32**:147-160

[2] Clough RW, Penzien J. Dynamics of Structures. 2nd ed. New York: McGraw-Hill; 1993

[3] Jiang Z, Christenson R. Experimental verification of an MR damper controlled highway bridge. 19th Analysis & Computation Specialty Conference, ASCE; 2010. pp. 347-358

[4] Sadiku S, Leipholz HHE. On the dynamics of elastic systems with moving concentrated masses. Ingenieur-Archiv. 1987;**57**:223-242

[5] Akin JE, Mofid M. Numerical solution for response of beams with moving mass. Journal of Structure Engineering, ASCE. 1989;**115**(1):120-131

[6] Pisal AY, Jangid RS. Vibration control of bridge subjected to multi-axle vehicle using multiple tuned mass friction dampers. International Journal of Advanced Structure Engineering. 2016;**8**:213-227

[7] Metin M, Guclu R. Rail vehicle vibrations control using parameters adaptive PID controller. Mathematical Problems in Engineering. 2014;**2014**, Article ID 728946, 10 p

[8] Metin M, Yücel ME, Paksoy M, Çetin Ş. Controlling rail vehicle vibrations using Magnetorheological damper. 3rd International Symposium on Railway Systems Engineering (ISERSE'16), October 13–15; 2016. Karabuk: Turkey

[9] Zong L, Gong X, Xuan S, Guo C. Semi-active H∞ control of high-speed railway vehicle suspension with magnetorheological dampers. Vehicle System Dynamics: International Journal of Vehicle Mechanics and Mobility. 2013;**51**(5):600-626

[10] Liao WH, Wang DH. Semiactive vibration control of train suspension systems via magnetorheological dampers. Journal of Intelligent Material Systems and Structures. 2003;**14**(3): 161-172

[11] Radeström S. Application of fluid viscous dampers to mitigate vibrations of high-speed railway bridges. International Journal of Rail Transportation. 2017;**5**(1):47-62

[12] Yang M. Longitudinal vibration control for a suspension bridge subjected to vehicle braking forces and earthquake excitations based on magnetorheological dampers. Journal of Vibration and Control; 2015;**22**(17):3659-3678

[13] Metin M. Raylı sistem araçlarının modellenmesi ve titreşimlerinin kontrolü", M.S. thesis. Istanbul: YTU; 2007

9

Wheel-Rail Impact by a Wheel Flat

Lin Jing

Abstract

The finite element method (FEM)-based wheel-rail rolling contact model with a fresh wheel flat was built to investigate the wheel-rail impact responses, where a comprehensive dynamic explicit algorithm was employed. Two basic dynamic effects (i.e., inertia effect and strain-rate effect) and temperature effect during the wheel-rail sliding process were considered. Influences of train speed, flat length and axle load on the wheel-rail impact responses were discussed in terms of wheel-rail impact force, von Mises equivalent stress, equivalent plastic strain and XY shear stress. Simulation results demonstrate that the FEM-based wheel-rail rolling contact model can well describe the strong nonlinearities in geometry, contact and material. The strain rate effect contributes to elevate the maximum von Mises equivalent stress and restrain the plastic deformation. The initial thermal stress can decrease the maximum von Mises equivalent stresses and maximum XY shear stresses, but can aggravate the plastic deformation. Furthermore, the flat-induced wheel-rail impact force, von Mises equivalent stress, equivalent plastic strain and XY shear stress are revealed to be sensitive to train speed, flat length and axle load.

Keywords: wheel flat, impact response, strain rate, finite element simulation, high-speed railway

1. Introduction

The wheel-rail interaction has become increasingly one of the most attractive and important topics in the field of rail transportation, since the serious wheel/rail failure will surely lead to a series of disasters [1, 2]. The wheel flat is a main type of potential dangerous factors of inducing the wheel/rail failure, which is usually generated by the two following factors: (i) the sudden lock of a running wheel during the braking process, resulting that the braking force exceeds temporarily the available wheel-rail friction force; and (ii) the sliding of the wheel on the rail under the circumstance of a local reduction of the wheel-rail adhesion force [3–5]. **Figure 1** shows a typical photograph of a wheel flat formed by wheel-rail sliding.

The consequences of flat-caused contact irregularity are to excite the significantly larger wheel-rail impact forces than the corresponding quasi-static wheel-loads. The resulted high impact forces may cause severe damages to both railway vehicles and track, including broken axles and axle boxes, damaged bearings, cracks penetrating into the wheels and rail fracture with the risk of derailment [6]. Besides, the presence of wheel flats will produce excessive noises and higher-frequency forced vibrations.

The wheel flat problem remains of great current research interest. This is because that the speeding-up and high-speed trains are expected to be gradually operated on the limit of the possible wheel-rail tractions in accelerating and decelerating, based on optimized and tight time schedules [5, 6]. Furthermore, there are two criteria for determining the allowable value of flat length at present [6]: one is the force-based criterion, which is based on the principle of impact load detectors in the track, and the other is the geometry-based criterion, which specifies an allowable length or depth (or a combination of both), of the flat spot. However, the force-based criterion does not match with the geometry-based criterion, attributed to the fact that there is no unique corresponding relationship between the flat length and the maximum wheel-rail impact force, especially for the high-speed case [5, 7]. Therefore, the wheel-rail impact problem due to the wheel flat needs to be further in-depth studied, so as to provide valuable management decisions, based on better understanding of corresponding regularities and mechanisms.

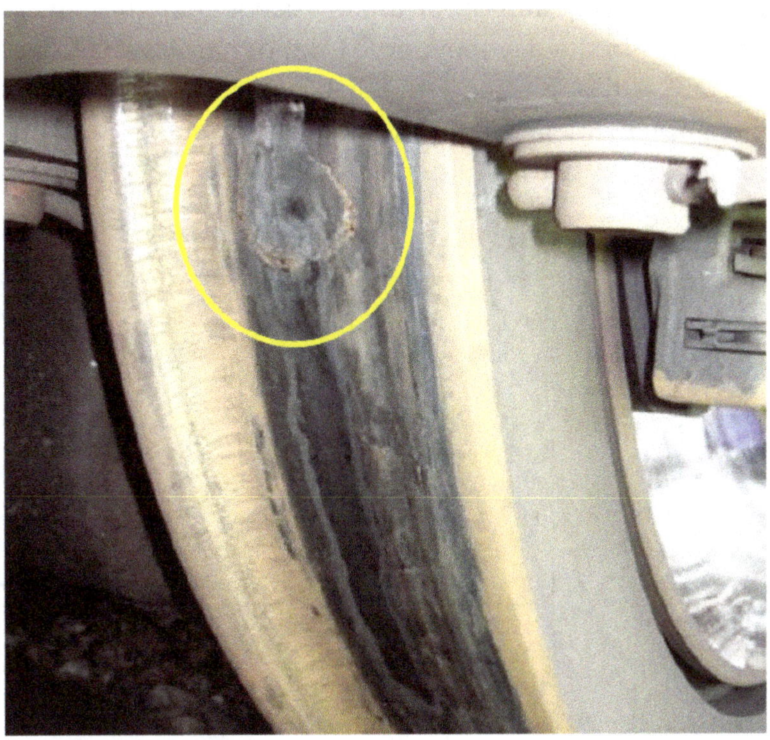

Figure 1. Photograph of a wheel flat formed by wheel-rail sliding [6].

2. Overview of wheel-rail impact problem

The wheel-rail impact problem induced by a wheel flat is usually simulated using the multi-body dynamics (MBD) approach [8, 9] and newly-developed finite element method (FEM) [10, 11]. Then the wheel flat is directly geometrically modeling or simulated by implementing a relative displacement excitation between the wheel and rail [8]. Considering a wheel with a flat rolling on the rail shown in **Figure 2 (a)** and **(b)**, the relative displacement excitations for the wheel flat can be represented by the vertical movement of the wheel center x_0 (positive downwards), which is given approximately by

$$x_0 = \begin{cases} z_0^2/2r & 0 \le z_0 \le l/2 \\ (l - z_0)^2/2r & l/2 < z_0 \le l \end{cases} \tag{1}$$

and

$$x_0 = \begin{cases} 4d\ (z_0/l)^2 & 0 \le z_0 \le l/2 \\ 4d\ ((l - z_0)/l)^2 & l/2 < z_0 \le l \end{cases} \tag{2}$$

for the fresh and rounded flats, respectively; where, $z_0 = r\theta$ is the longitudinal position of the wheel center, and r is the radius of the wheel; l and d is the length and depth of the flat, respectively.

The MBD approaches are well-qualified for the low-frequency dynamics analyses, but they are not suitable for the high-frequency (beyond 20 Hz) dynamics analyses due to the assumptions of rigid bodies [11]. Although the dynamic contact forces were also calculated by the MBD approach with implemented the rolling contact models [12, 13], most of the rolling

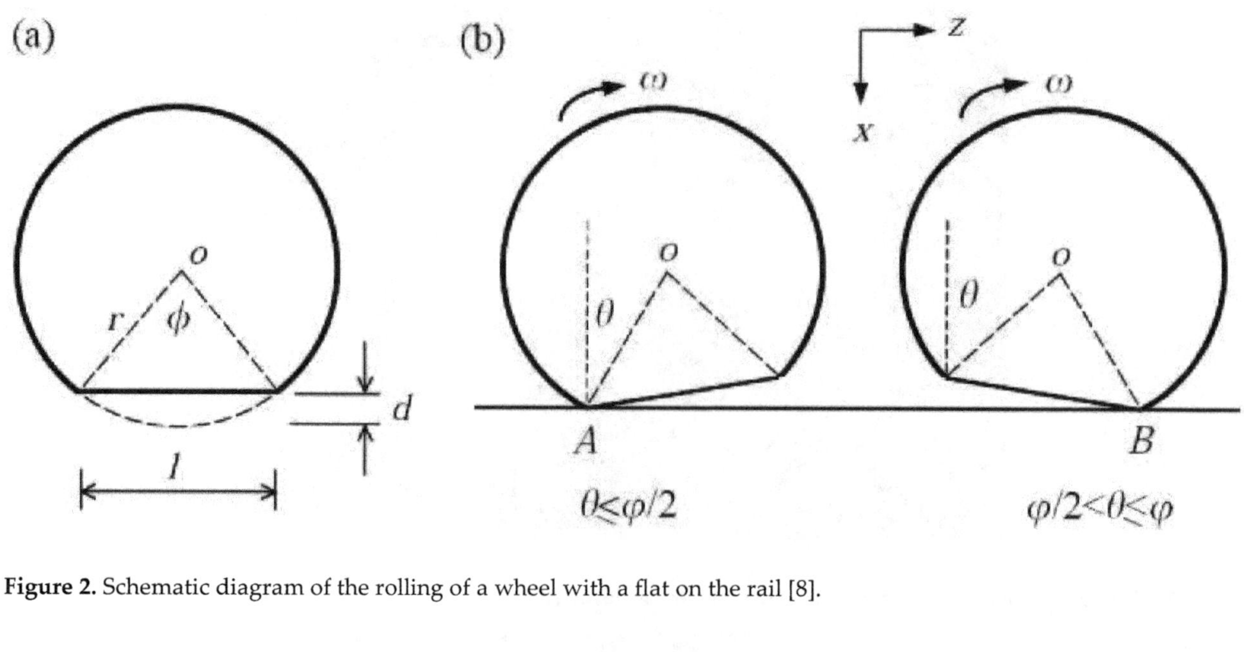

Figure 2. Schematic diagram of the rolling of a wheel with a flat on the rail [8].

contact theories were based on the Hertzian contact and steady rolling assumptions. These assumptions are difficult to describe accurately the wheel-rail impact behavior due to its non-linearities of materials and the dynamic contact. Besides, the MBD approaches are failed to solve directly the wheel-rail contact stress and strain states, and the dynamic effects are usually ignored to a large extent. Nevertheless, the FEM-based wheel-rail rolling contact simulations have abilities of representing the actual geometrical and kinematic characteristics of the wheel-rail system, and considering the strong nonlinearities in geometry, contact and material [5, 7]. The FEM is therefore more widely-used in wheel-rail impact simulations. It should be noted that, under the high-speed condition, the inertia effect cannot be neglected and the strain rate effect of materials becomes more and more important then. Besides, the thermal stress induced by induced by the friction temperature rising during the wheel-rail sliding process will play a more important role in the wheel-rail interaction.

3. FEM-based wheel-rail rolling contact model

The entire finite element model of the wheel-rail system, built by using commercial software Hypermesh, consists of two wheels (the one has a fresh flat), one axle and two rails. The wheel with the radius of 430 mm has a S1002CN tread, while the rail with a base slope of 1: 40 is the 60 kg/m rail. The whole structure except the substructure below the rail of the wheel-rail system was modeled, and a typical 3-D wheel-rail rolling contact model with a flat length of 40 mm was presented in **Figure 3**, where the width of the flat is equal to 25 mm. To balance the computational precision and efficiency, the contact area of the wheel flat was fined meshed with the size of 4 mm × 4 mm, while the other areas were medially meshed. The entire model was mesh into the 8-node solid element, and comprises 516,629 nodes and 479,038 elements.

The mechanical behaviors of all the wheel-rail components were described by a plastic kinematic hardening constitutive model *MAT_PLASTIC_KINEMATIC [5, 7], where mechanical parameters used in the rolling contact simulations were listed in **Table 1**. Nodal all-DOFs (degree of freedom) constraints were imposed to the bottom of rails, to represent the clamped

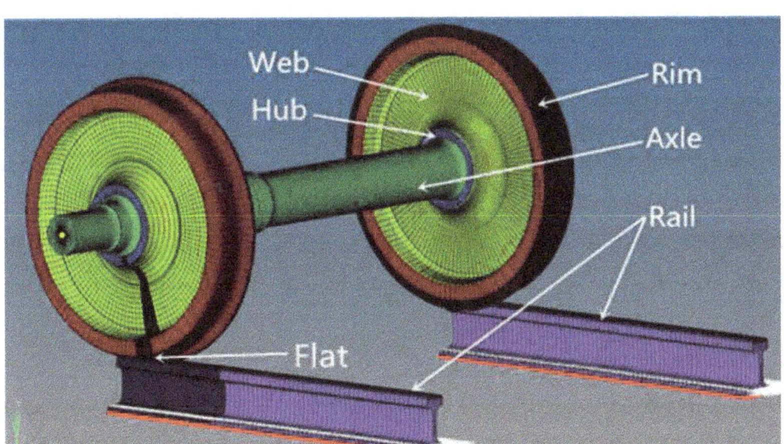

Figure 3. FE model of the wheel-rail system with a 40 mm flat.

Component	Elastic modulus (GPa)	Density (kg/m³)	Poisson ratio	Yield stress (MPa)	Tangent modulus (GPa)
Rim	213	7800	0.3	561	21
Web	216	7800	0.3	395	21
Hub	213	7800	0.3	417	21
Axle	206	7800	0.3	560	21
Rail	193	7800	0.3	525	19

Table 1. Material parameters used in the rolling contact simulations.

boundary condition, and an axial translational constraint was set to all nodes of the axle. The gravitational acceleration was considered and endowed to all the wheel-rail components. Based on the criterion of EN13104 [14], the static wheel-loads were equivalent to two concentrated forces applied to the two ends of the axle. The initial velocity of the wheel set was endowed by an equivalent translational velocity for both wheels and the axle, and a corresponding rotational velocity only for wheels. An automatic surface-to-surface contact option with the penalty method was generally used for the whole wheel-rail system, to solve the wheel-rail rolling contact.

4. Wheel-rail impact simulation: inertia effect

The whole dynamic wheel-rail impact response process was simulated using finite element code LS-DYNA 3D explicit algorithm, where the inertia effect is considered automatically by the following dynamic equilibrium equation [15].

$$[M]\{\ddot{u}\} + [C]\{\dot{u}\} + [K]\{u\} = \{F\} \tag{3}$$

where **[M]**, **[C]** and **[K]** are the structural mass, damping and stiffness matrixes, respectively; $\{u\}$ and $\{F\}$ are the nodal displacement and applied load vectors, respectively.

The maximum vertical wheel-rail impact forces induced by a wheel flat are plotted in **Figure 4** as a function of train speed and flat length. For a given flat length, the non-monotonic relationships between the maximum vertical impact force and train speed are clearly presented, and the peak values of maximum vertical impact forces occur at train speed of 150 km/h. This non-monotonic relationship between the peak vertical force and train speed may be related with the loss of contact under different speeds. For a given train speed, the maximum vertical impact force non-linearly increases with the flat length, and the influence of flat length on the maximum vertical impact force seems to be significant within the speed range from 150 to 250 km/h. For a given flat length and train speed, the influence of axle load on the wheel-rail impact response can be also examined. The maximum vertical impact forces are increased with the axle load for each flat length case.

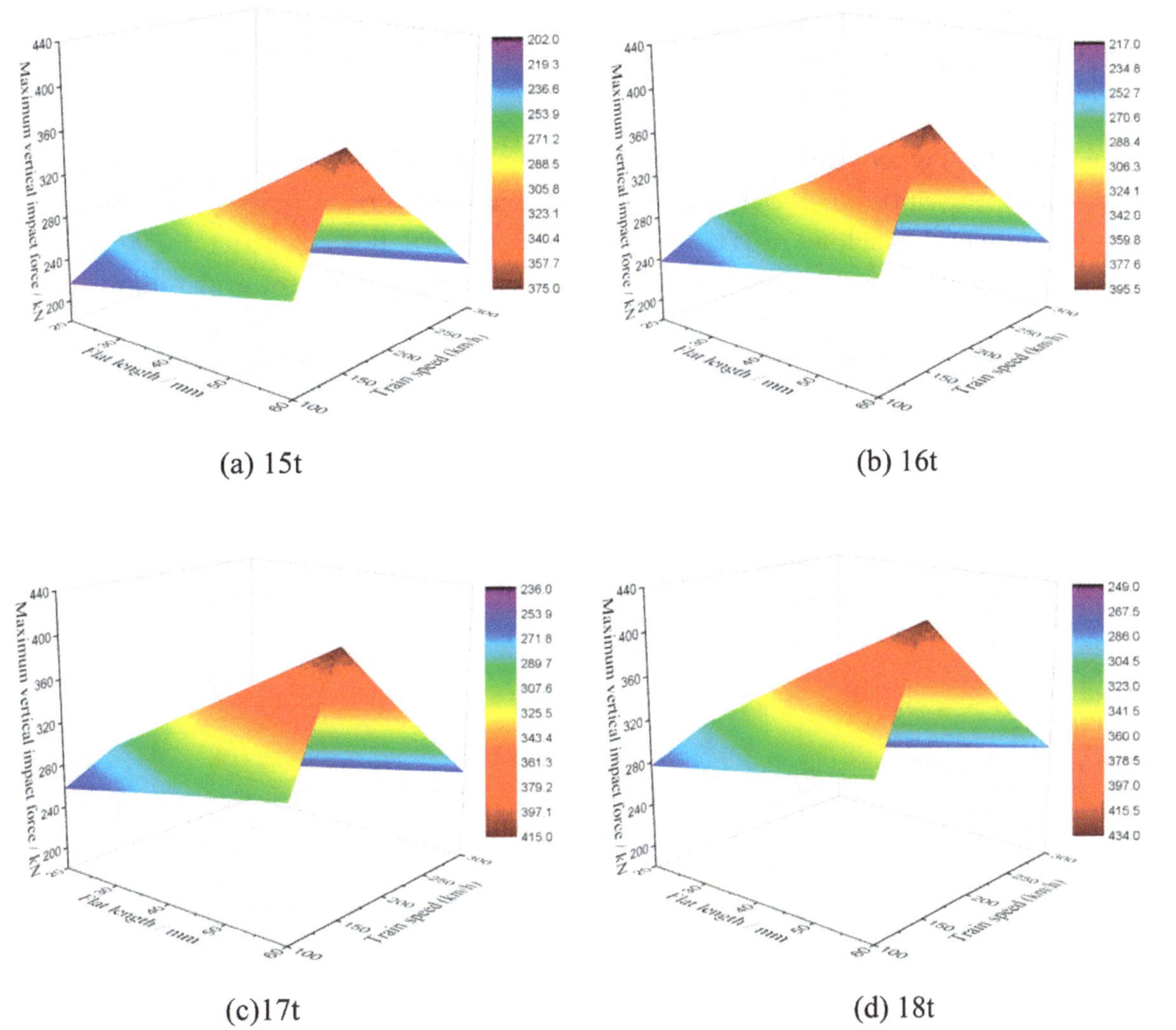

Figure 4. Maximum flat-induced wheel-rail impact forces as a function of train speed and flat length. (a) 15t, (b) 16t, (c) 17t and (d) 18t.

5. Wheel-rail impact simulation: inertia effect + strain rate effect

As stated earlier, the strain rate effect of wheel/rail materials should be taken into account in wheel-rail impact simulations with the continuous raising of train speed. Thus, the strain rate effect is considered in material models of LS-DYNA via inputting the corresponding strain rate-dependent parameters, which are obtained from the quasi-static compressive and split Hopkinson pressure bar tests at a wide range of strain rates. In the present case, the Cowper-Symonds model which scales the yield stress by a following strain rate dependent factor is employed.

$$1+\left(\frac{\dot{\varepsilon}}{C}\right)^{1/P} \tag{4}$$

where $\dot{\varepsilon}$ is the strain rate, C and P are two strain rate parameters, corresponds to the strain rate parameter,C, (SRC) and strain rate parameter,P, (SRP) options in the material model, respectively. Based on the experimental fitting [16, 17], the strain rate sensitive material parameters in the present study are set as: C = 45,635 s^{-1} and P = 3.21 for the rim steel and C = 1733 s^{-1} and P = 0.30 for the rail steel, respectively.

The typical vertical wheel-rail impact force history curves, obtained from simulations with or without strain rate-dependent parameters, for the condition of train speed of 200 km/h, flat length of 40 mm and axle load of 17t, are plotted in **Figure 5**. It is clear that the vertical wheel-rail impact force responses derived from both strain rate-dependent and rate-independent

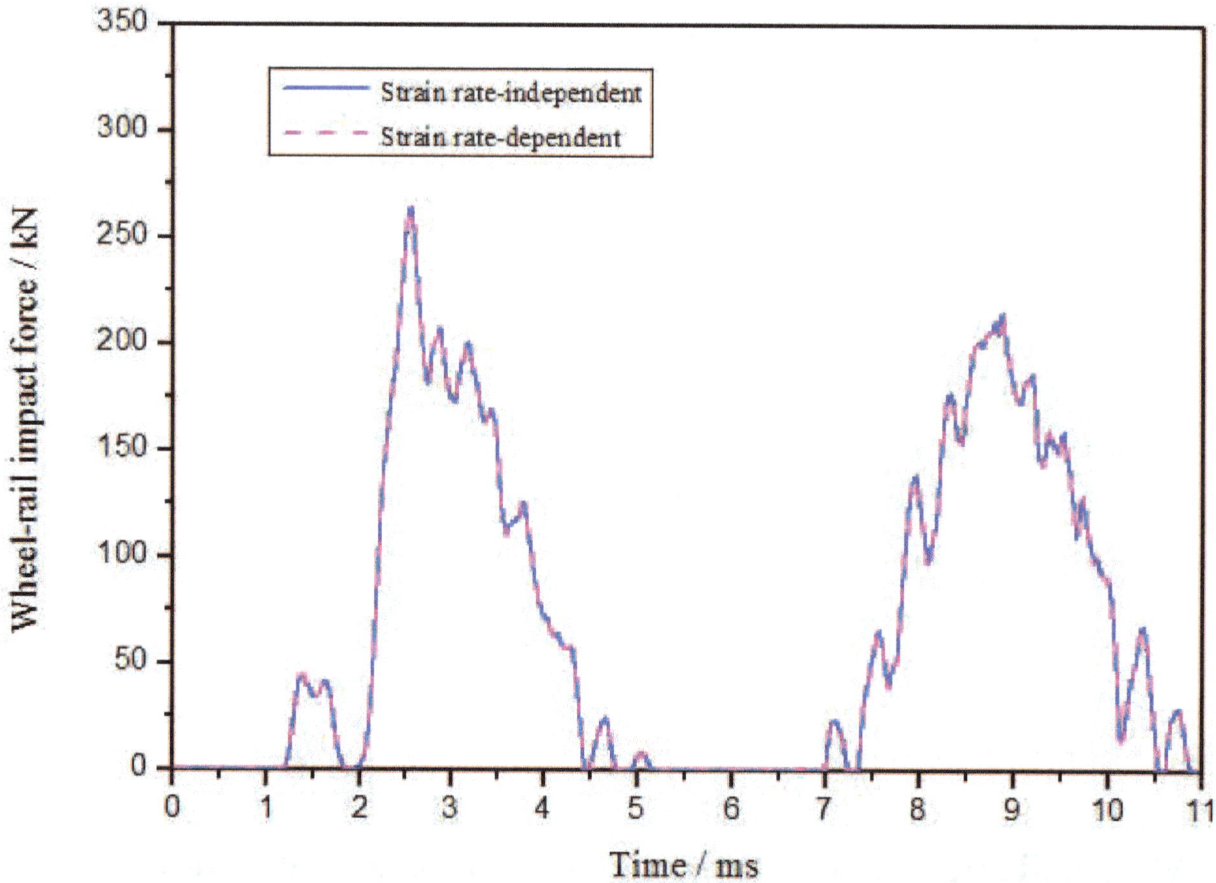

Figure 5. Wheel-rail impact force history curves (17t axle load, 200 km/h train speed and 40 mm flat).

simulations are coincident. This is to say, the strain rate effect of wheel/rail materials has no influence on the vertical impact force response. The first peak impact force with the amplitude of 265 kN, which is generated by the impact of wheel flat against the rail, occurs at 2.6 ms. This impact load is carried directly by the rail, especially for the wheel-rail contact area, since the impact force had no time to transfer to the infrastructure below the rail during the very short time. As a result, the large impact load may induce the plastic deformation and fatigue failure of the wheel and rail.

The typical von Mises equivalent stress, XY shear stress and equivalent plastic strain responses during the wheel-rail impact process derived from both rate-dependent and rate-independent simulations are plotted in **Figures 6–8**, respectively. It is seen from **Figures 6** and **7** that

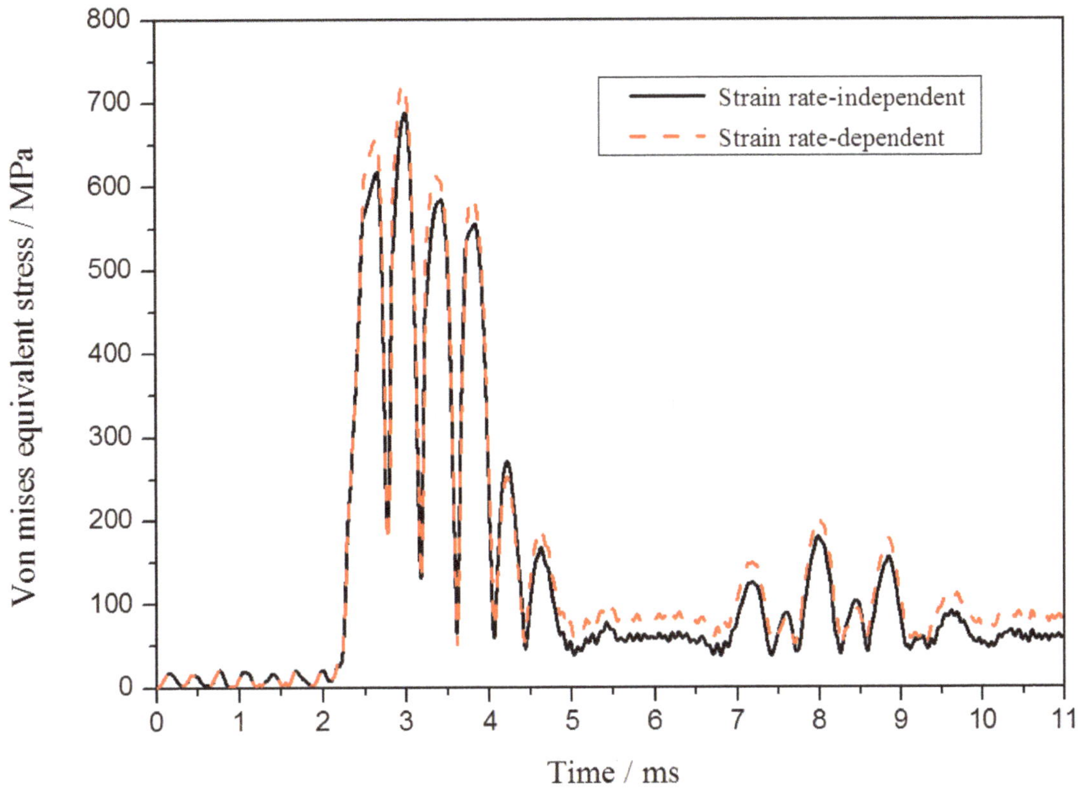

Figure 6. Von Mises equivalent stress history curves (17t axle load, 200 km/h train speed and 40 mm flat).

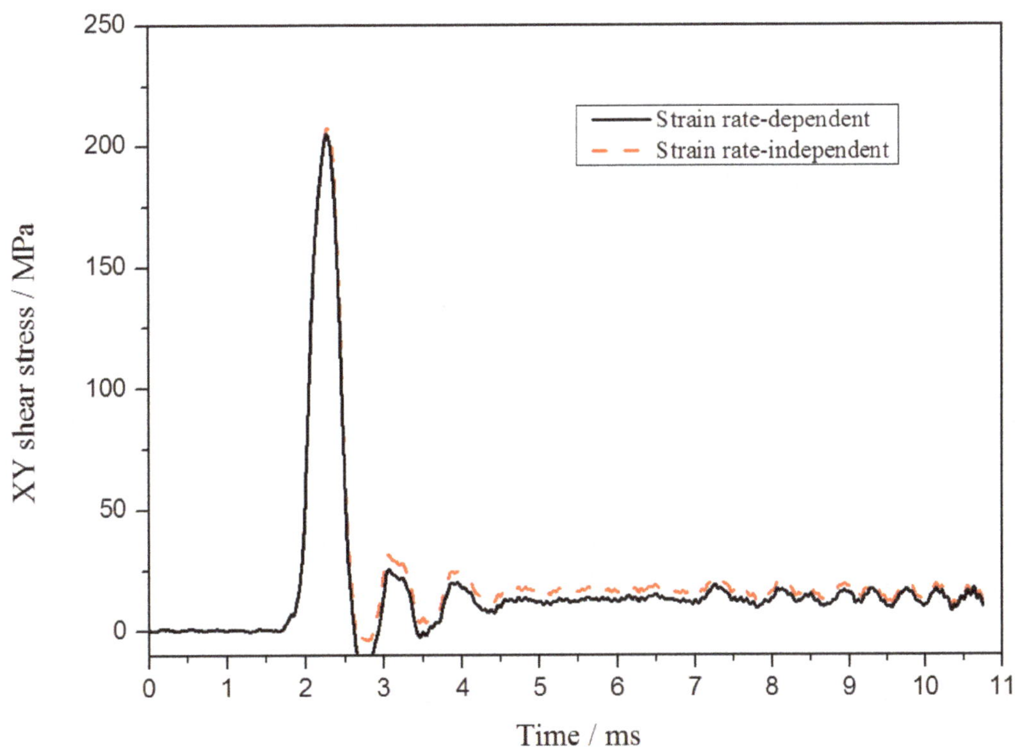

Figure 7. XY shear stress versus time curves (17t axle load, 200 km/h train speed and 40 mm flat).

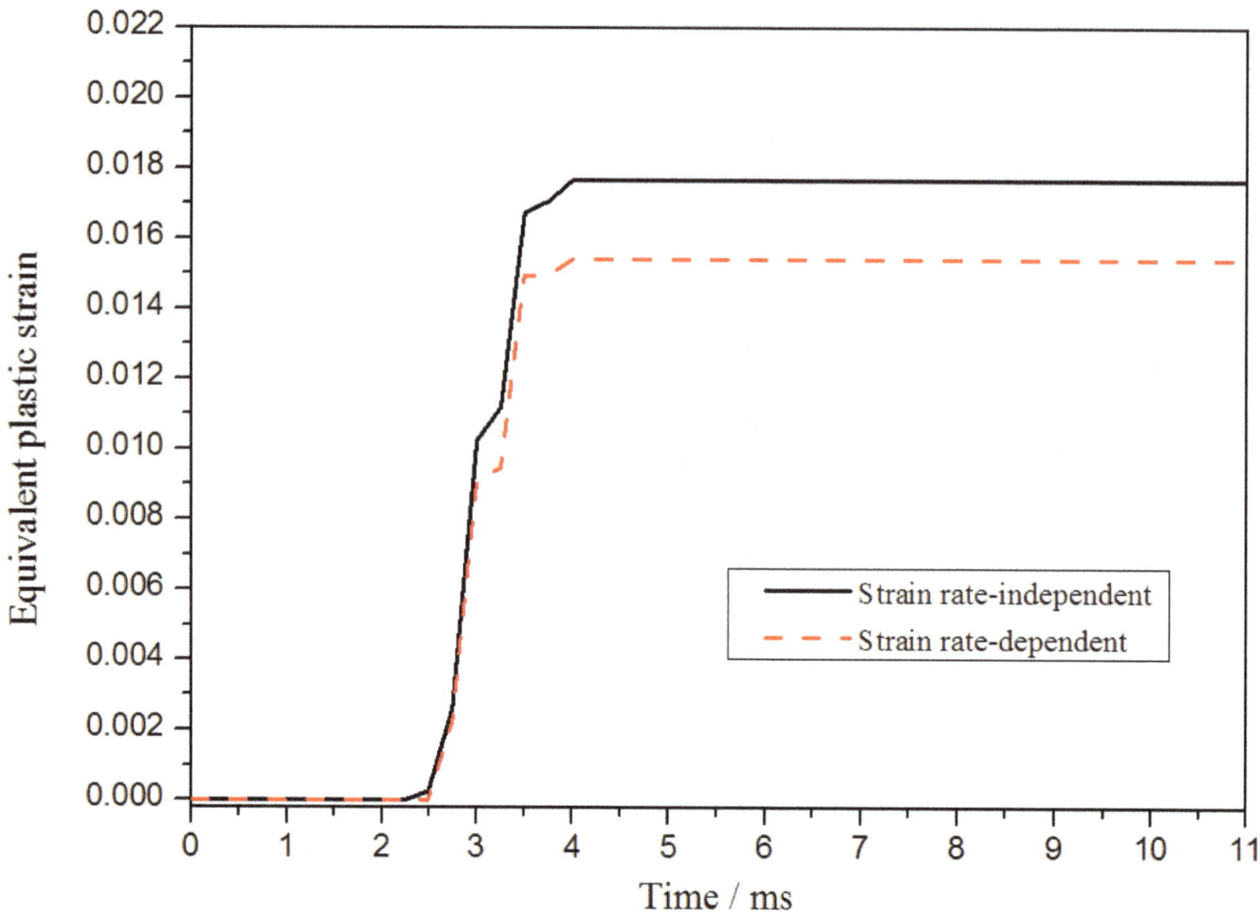

Figure 8. Equivalent plastic strain history curves (17t axle load, 200 km/h train speed and 40 mm flat).

both the von Mises equivalent stress and XY shear stress are significantly larger when the strain rate effect of wheel/rail materials is considered, while from **Figure 8** that the equivalent plastic strain gives an obviously lower value for the strain rate-dependent simulations. These indicate that the von Mises equivalent stress, shear stress and equivalent plastic strain are sensitive to the strain rate, and the hardening effect of the strain rate results the higher von Mises equivalent stress and shear stress, but the lower equivalent plastic strain.

6. Wheel-rail impact simulation: dynamic effects + temperature effect

In this section, the friction temperature rising during the wheel-rail sliding process is also considered in the wheel-rail impact simulation, together with inertia effect and strain rate effect. Thermal stress fields of the wheel and rail, derived from the thermal analysis [5], are inputted to the wheel-rail rolling contact finite element model as an initial pre-stress.

6.1. Typical characteristics of wheel-rail impact responses

Figures 9–11 present a typical set of wheel-rail impact force, von Mises equivalent stress and equivalent plastic strain history curves for the condition of train speed of 200 km/h, flat length

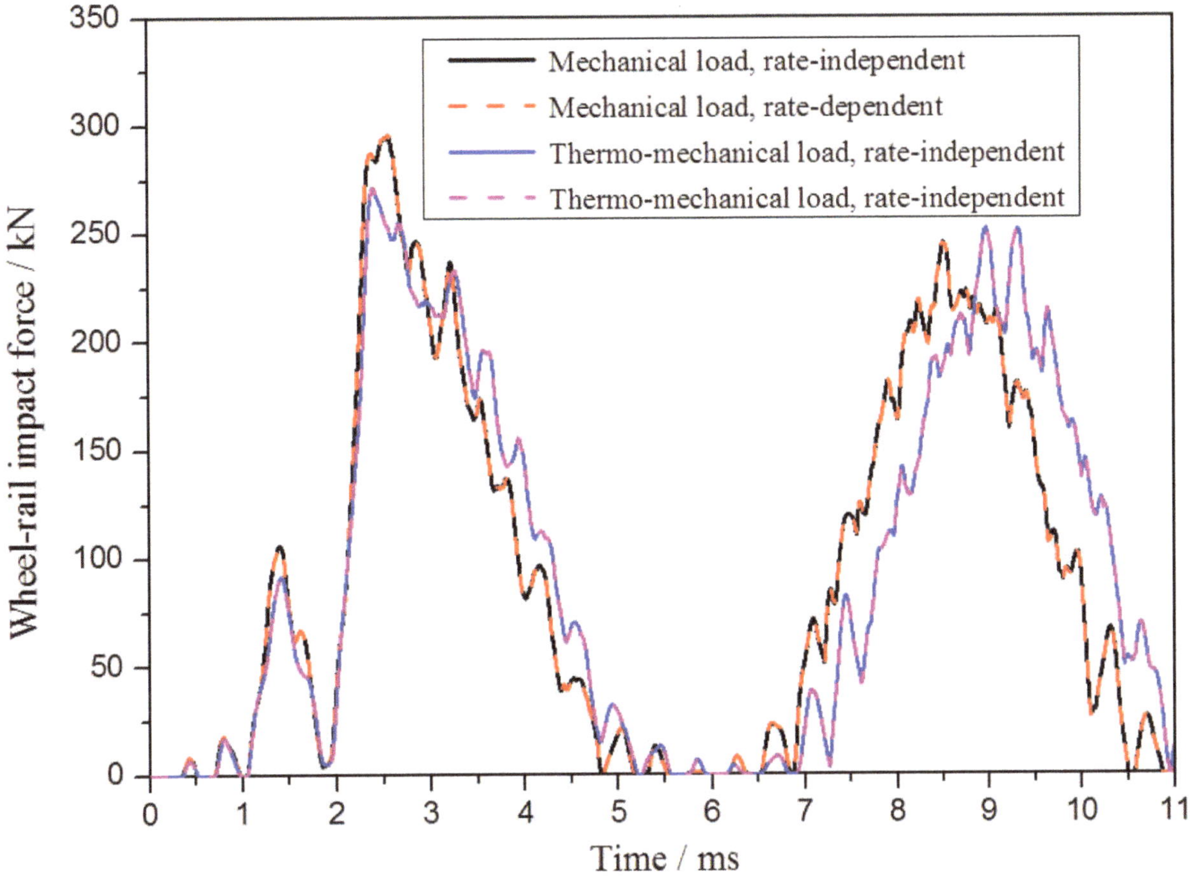

Figure 9. Wheel-rail impact force history curves under mechanical and thermo-mechanical loads.

of 40 mm and axle load of 17t, under mechanical and thermo-mechanical loads, respectively. It is seen from **Figure 9** that the starting times (at ~1.9 ms) and the moments corresponding to the peak forces (at ~2.5 ms) of the wheel-rail impact induced by a flat under four conditions are almost the same, in other words, both quantities are regardless of the thermal load (stress) and strain rate. However, after the occurrence of the wheel-rail impact, the initial thermal stress contributes to prolong the duration of wheel-rail impact and the duration of the second dropping of the wheel, due to the weakening of the stiffness of both wheel and rail caused by the initial thermal stresses. Besides, the wheel-rail impact force responses under either mechanical or thermo-mechanical loads are independent with the strain rate. However, the peak values of the wheel-rail impact force under thermo-mechanical load are slightly lower than those under pure mechanical load. This is the consequent of the increasing impact duration under thermo-mechanical load for a given initial momentum of the wheel.

From **Figure 10**, the change tendencies of the von Mises equivalent stress history curves under four conditions are similar, but the peak values present certain sensitivity to the initial thermal load (stress) and strain rate. The peak von Mises equivalent stress under pure mechanical load with the strain rate considered is the largest, and then that under pure mechanical load without considering strain rate, following by that under thermo-mechanical load with considering the strain rate, and that under thermo-mechanical load without considering strain rate is the smallest. Obviously, the superposition of the initial thermal stress and mechanical stress

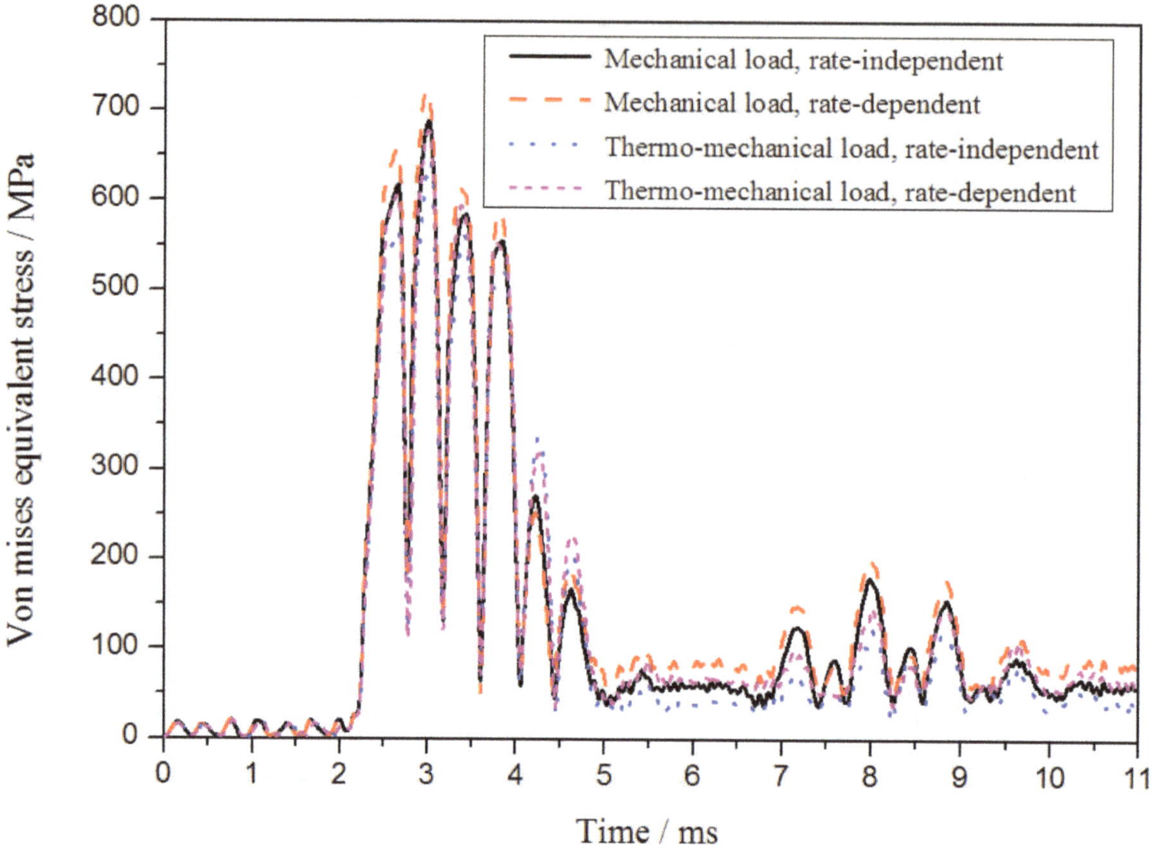

Figure 10. Von Mises equivalent stress history curves under mechanical and thermo-mechanical loads.

weaken the load-carrying capability of wheel/rail materials under thermo-mechanical load, resulting in a lower von Mises equivalent stress value than that under pure mechanical load.

The accumulated equivalent plastic strains for four various studied conditions are plotted as the curves in **Figure 11**. The maximum equivalent plastic strains are occurred at the wheel tread for all conditions, and with the largest value of 2.02%. The initial thermal stress contributes to exacerbate the plastic deformation of the wheel tread, while the strain rate hardening effect of wheel/rail materials will restrain the plastic deformation.

6.2. Influence of train speed on wheel-rail impact responses

The maximum wheel-rail impact forces induced by two lengths of flat (i.e., 40 mm and 60 mm) under mechanical and thermo-mechanical loads are plotted in **Figure 12** as a function of train speed. The maximum wheel-rail impact forces are shown to be insensitive to the strain rate of wheel/rail materials under the same loading condition for each train speed. However, the maximum wheel-rail impact forces under thermo-mechanical load are less than those values under pure mechanical load, regardless of the strain rate; this is attributed to the initial thermal stress, as stated earlier. Besides, the maximum wheel-rail impact forces first increase with the increasing train speed until reaching the peak value at the speed of 150 km/h, and then decrease with the increasing train speed for the train speed beyond

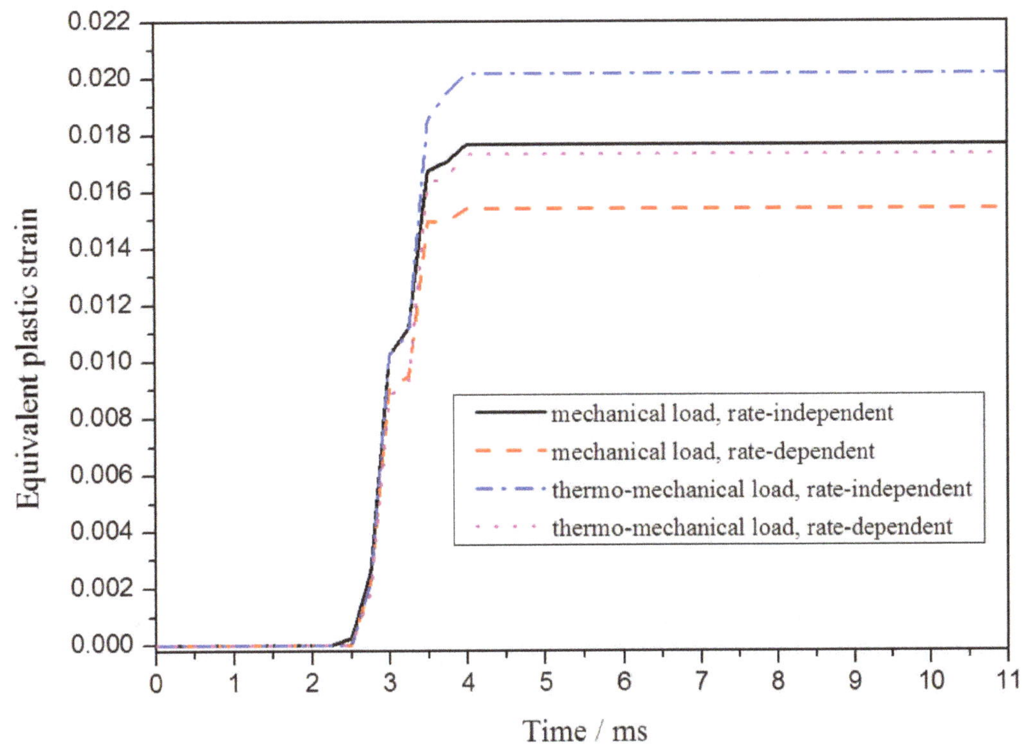

Figure 11. Equivalent plastic strain history curves under mechanical and thermo-mechanical loads.

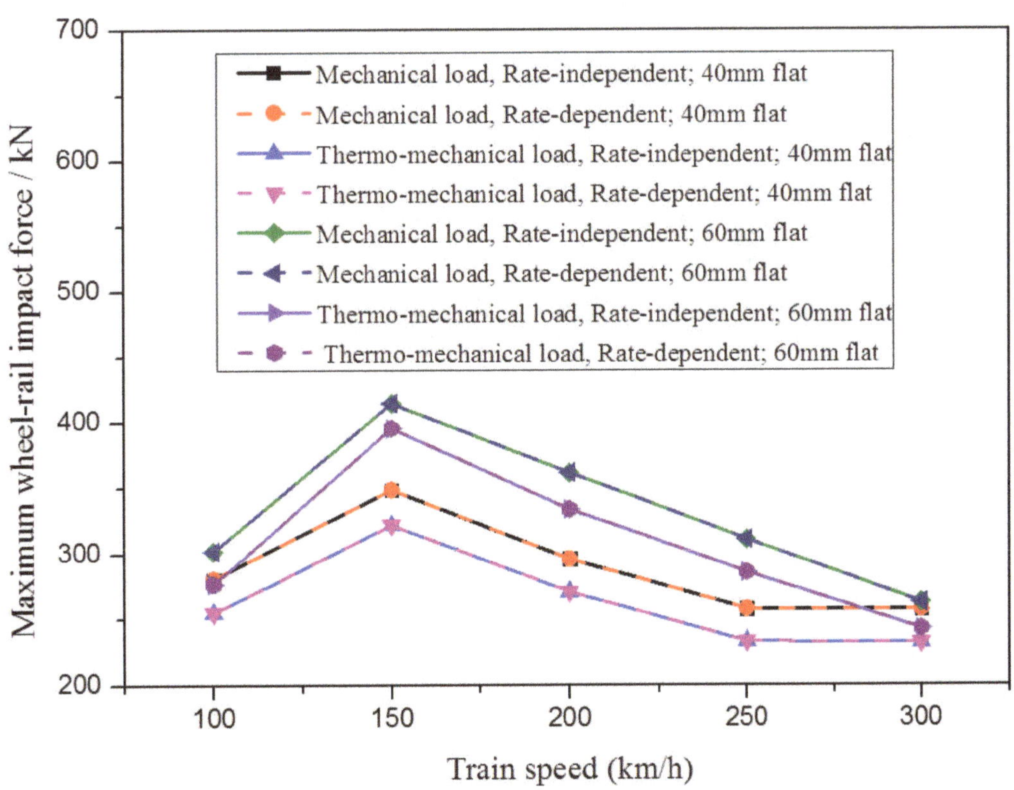

Figure 12. The maximum wheel-rail impact forces as a function of train speed.

150 km/h. This phenomenon may be explained reasonably with the dynamic contact loss opinion [2, 8]: when the wheel flat impacted against the rail, the impulse of the wheel rapidly transferred to the momentums of the wheel and rail; then the wheel tended to move upwards while the rail tended to move downwards, resulting in the loss of dynamic contact. For the lower train speed, the initial impulse of the wheel is low, and the momentums of the wheel and rail are also low; the quasi-static wheel load can maintain the wheel-rail contact, so the wheel-rail impact force increases with the train speed. However, for the higher speed case, the large impulse induces the large momentums of the wheel and rail; during the downward movement of the rail, the wheel still keeps falling for a while because of its large inertia, so the maximum impact force is attenuated, as a result of the reduction of the dynamic contact.

Figures 13–15 show the relationships of maximum von Mises equivalent stress, maximum equivalent plastic strain and maximum XY shear stress of the wheel with train speed, respectively. As shown in **Figure 13**, the maximum von Mises equivalent stress presents a similar change tendency versus train speed with that of the wheel-rail impact force, but it presents large strain rate sensitivity (i.e., the strain rate hardening effect elevates the maximum von Mises equivalent stress values under all train speed conditions). Under thermo-mechanical load, the maximum von Mises equivalent stresses are obviously greater than those under pure mechanical load. In **Figure 14**, the maximum equivalent plastic strain first increases and then decreases with the increasing train speed, and the corresponding peak value is occurred at the train speed of 200 km/h, for each loading condition. The strain rate effect of wheel/rail materials contributes to restrain the plastic deformation, resulting in a lower equivalent plastic strain value. It is interesting that the maximum equivalent plastic strains

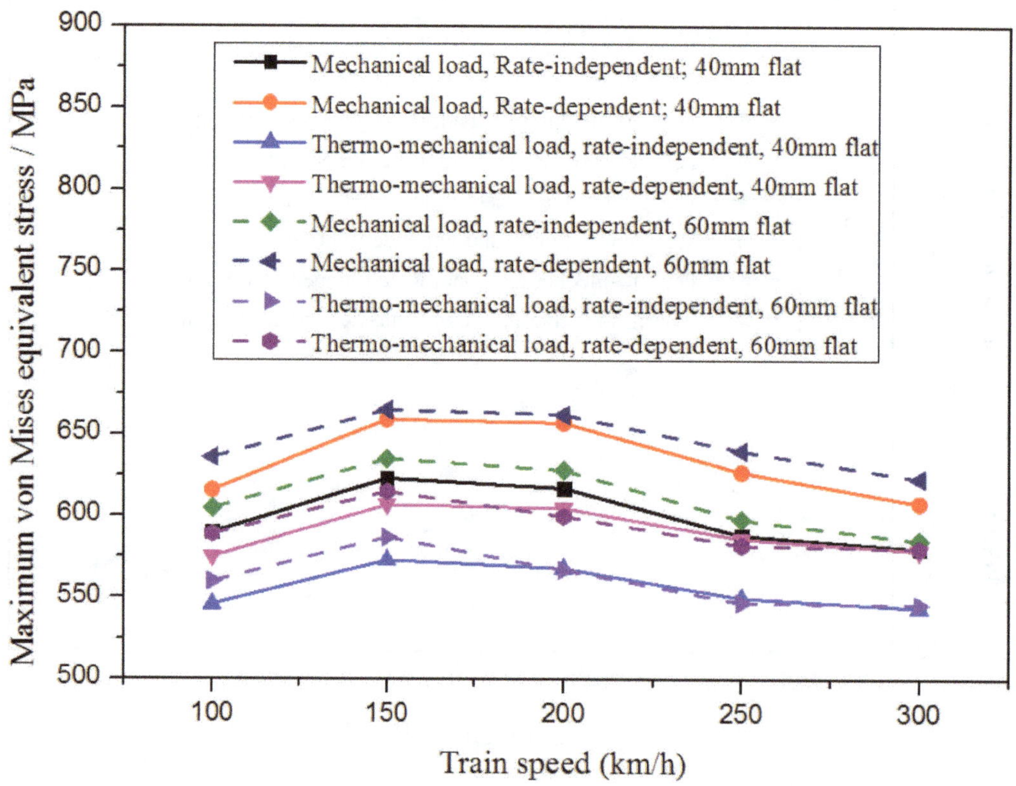

Figure 13. Maximum von Mises equivalent stresses as a function of train speed.

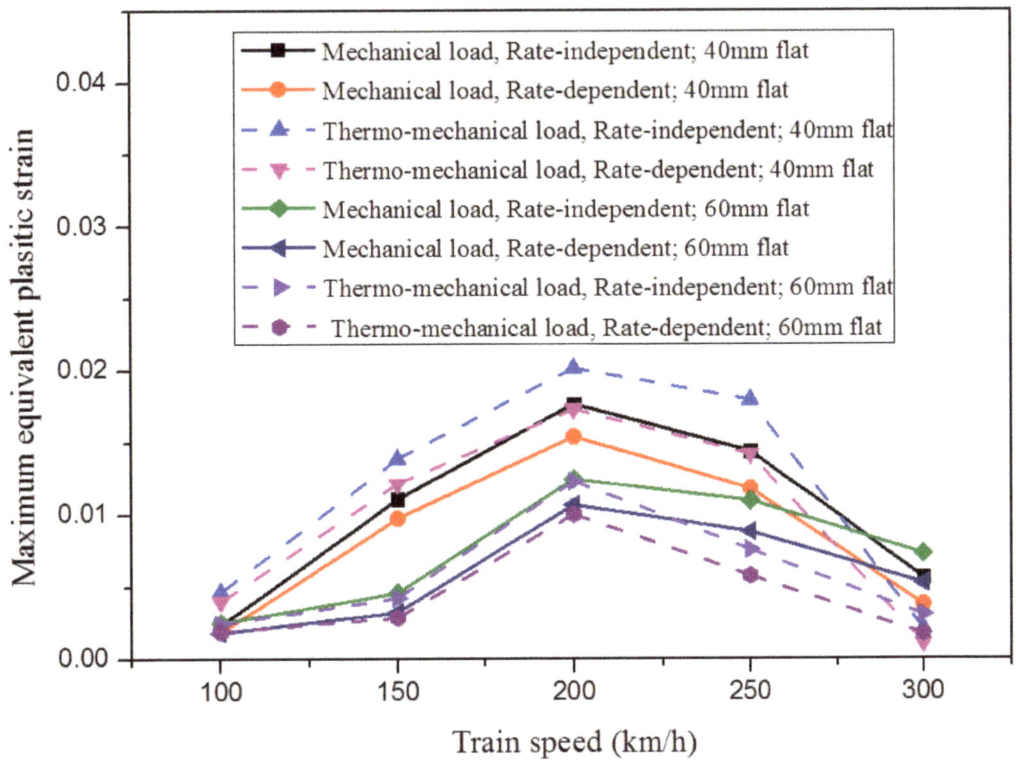

Figure 14. Maximum equivalent plastic strains as a function of train speed.

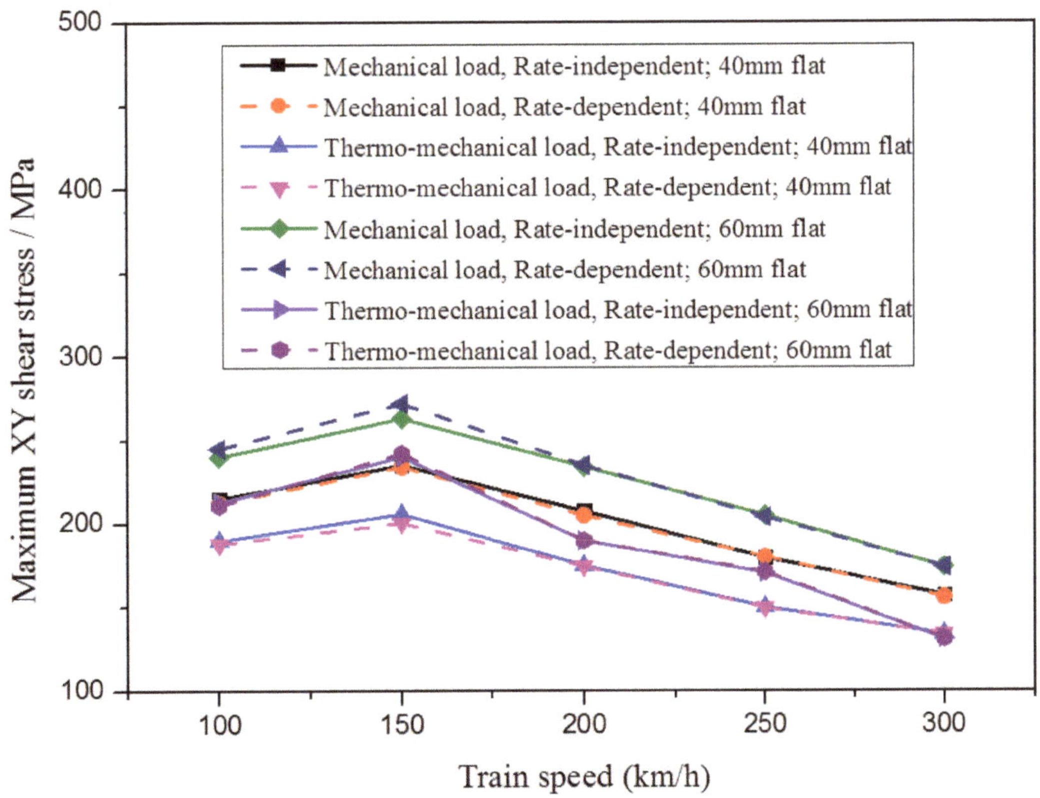

Figure 15. Maximum XY shear stresses as a function of train speed.

under thermo-mechanical load are less than those under pure mechanical load at the speed of 300 km/h. This abnormal tendency indicates that the impact position of the wheel flat has changed obviously under the speed of 300 km/h, since the equivalent plastic strain is not only related to the impact force but also the impact position. It can be found from **Figure 15** that the maximum XY shear stresses present weak strain rate sensitivity, especially with the increase of train speed; and they are lower under thermo-mechanical load than those under mechanical load for each speed condition.

6.3. Influence of flat length on wheel-rail impact responses

Based on **Figures 12–15**, the influence of flat length on the wheel-rail impact responses can be also identified. The maximum wheel-rail impact forces for each condition are revealed to be increased with the increasing the flat length. Similarly, the strain rate effect contributes to elevating the maximum von Mises equivalent stress, restraining the plastic deformation and has no effect for the XY shear stress at each condition; the initial thermal stress can decrease the maximum von Mises equivalent stresses and maximum XY shear stresses, and aggravate the plastic deformation. With the increase of flat length, the maximum von Mises equivalent stresses and the maximum XY shear stresses increase, and the maximum equivalent plastic strains decrease correspondingly.

6.4. Influence of axle load on wheel-rail impact responses

For a given train speed of 200 km/h and flat length of 60 mm, five different axle loads (i.e., 15t, 16t, 17t, 18t and 19t) are imposed to the wheel-rail contact model, to explore the influence of axle load. The maximum wheel-rail impact forces are increase approximately linearly with

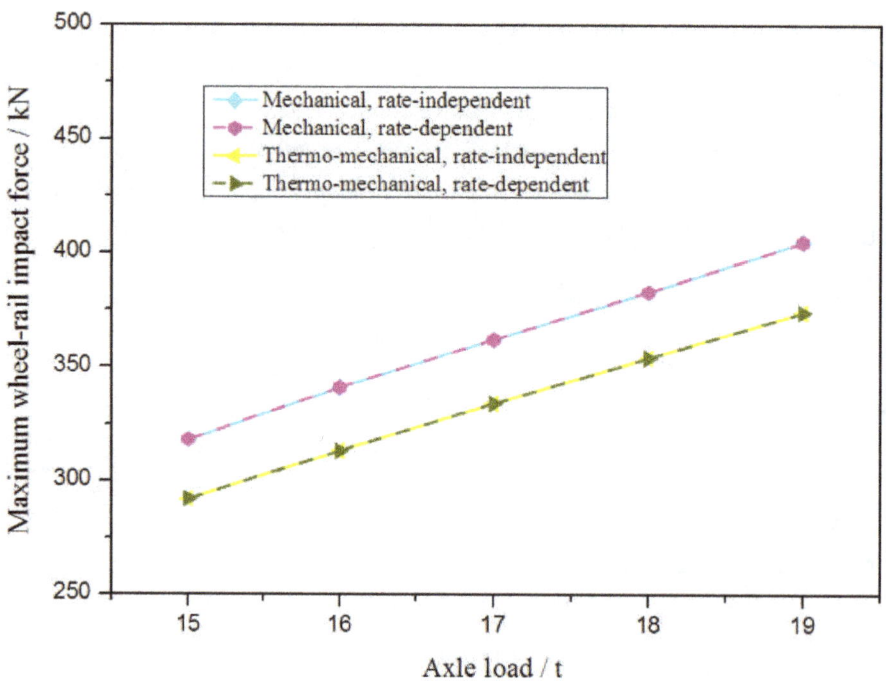

Figure 16. Maximum wheel-rail impact forces as a function of axle load.

the increasing axle load, and the values under thermo-mechanical load are lower than those under mechanical load, as shown in **Figure 16**.

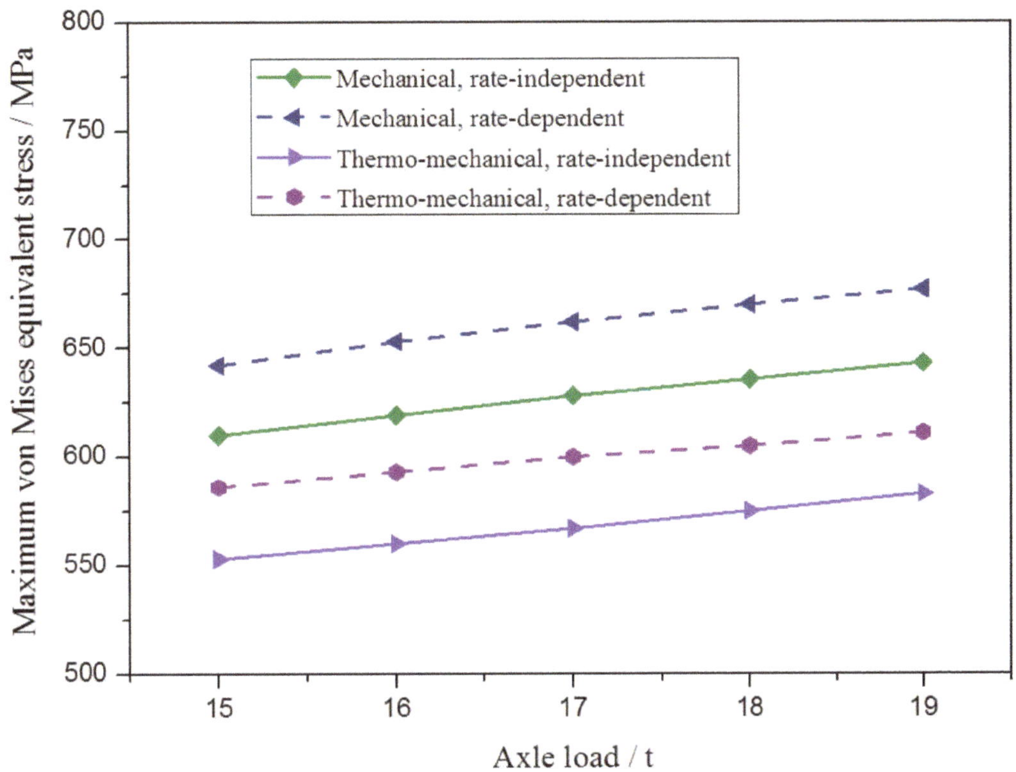

Figure 17. Maximum von Mises equivalent stresses as a function of axle load.

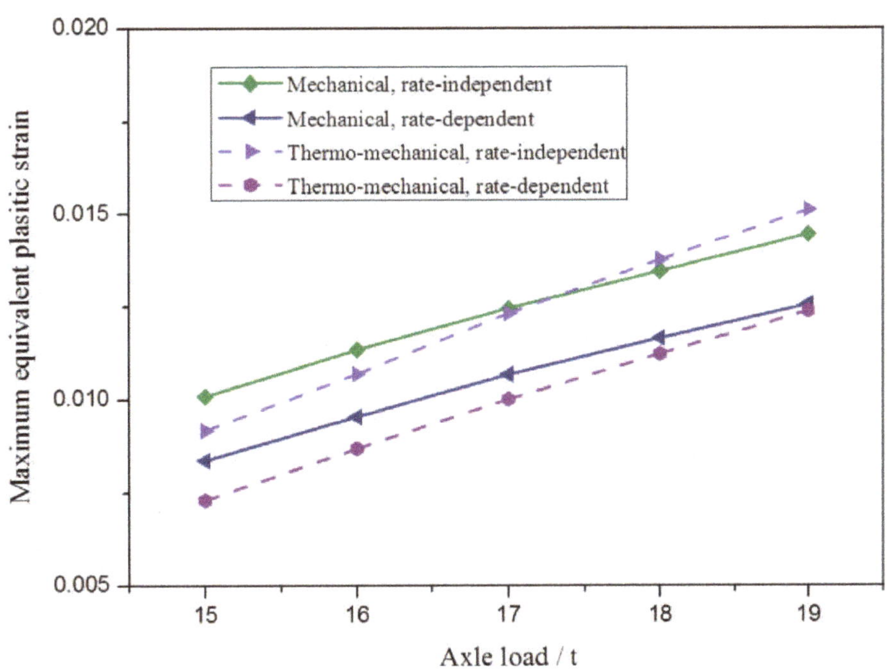

Figure 18. Maximum equivalent plastic strains as a function of axle load.

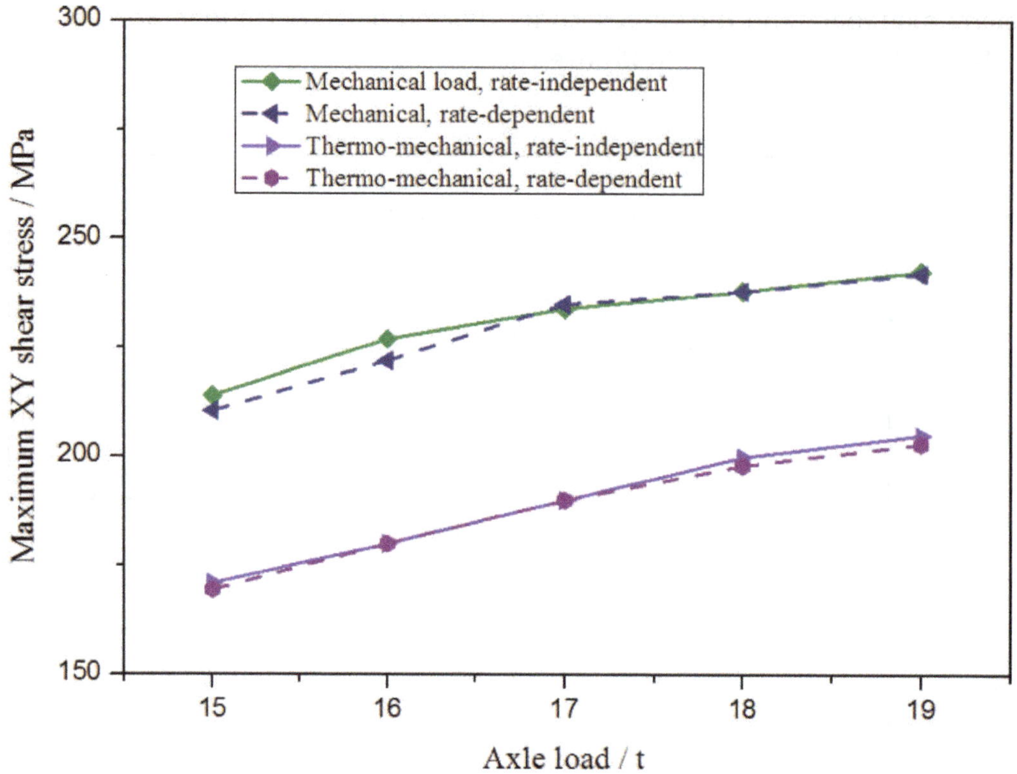

Figure 19. Maximum XY shear stresses as a function of axle load.

The maximum von Mises equivalent stress and maximum equivalent plastic strain are increased approximately linearly with the increasing axle load, while the maximum XY shear stress is increased with the axle load, as shown in **Figures 17–19**. The maximum von Mises equivalent stress increases significantly and the maximum equivalent plastic strain decreases correspondingly for each axle load case, as the strain rate effect considered. However, the strain rate has almost no influence on the maximum XY shear stress for all axle load conditions.

7. Conclusions

A 3-D wheel-rail rolling contact finite element model with a fresh wheel flat is built, and a comprehensive dynamic simulation method is used to investigate the wheel-rail impact responses. Inertia effect, strain-rate effect and temperature effect are included in the present simulation. Influences of train speed, flat length and axle load on the flat-induced wheel-rail impact responses were discussed in term of wheel-rail impact force, von Mises equivalent stress, equivalent plastic strain and XY shear stress. Some main conclusions can be drawn out:

1. The FEM-based wheel-rail rolling contact simulation can well characterise the actual geometrical and kinematic characteristics of the wheel-rail system, and well describe the strong nonlinearities in geometry, contact and material.
2. With the continuous raising of train speed, the structural inertia effect and strain rate effect of materials cannot be ignored in wheel-rail impact simulations, the thermal stress induced

by induced by the friction temperature rising during the wheel-rail sliding process also play a more important role in the wheel-rail interaction.

3. The strain rate hardening effect contributes to elevate the von Mises equivalent stress and restrain the plastic deformation; and the initial thermal stress due to the sliding friction will aggravate the plastic deformation of wheel and rail.

4. The wheel-rail impact responses first increases and then decreases with the train speed; however, they are increased with the increasing axle load. With the increase of flat length, the maximum impact force, von Mises equivalent stresses and XY shear stresses increase, and the maximum equivalent plastic strains decrease correspondingly.

Acknowledgements

The authors greatly appreciate the financial support by the National Natural Science Foundation of China (Grant no. 51475392), the Fundamental Research Funds for the Central Universities (Grant no. 2682015RC09) and the Research Fund of State Key Laboratory of Traction Power (Grant no. 2015TPL_T02).

Author details

Lin Jing

Address all correspondence to: jinglin_426@163.com; jinglin@home.swjtu.edu.cn

State Key Laboratory of Traction Power, Southwest Jiaotong University, Chengdu, P.R. China

References

[1] Newton SG, Clark RA. An investigation into the dynamic effects on the track of the wheel flats on railway vehicles. Journal of Mechanical Engineering Science. 1979;**21**(4):287-297

[2] Pieringer A, Kropp W, Nielsen JCO. The influence of contact modelling on simulated wheel/rail interaction due to wheel flats. Wear. 2014;**314**(1):273-281

[3] Steenbergen MJMM. The role of the contact geometry in wheel–rail impact due to wheel flats. Vehicle System Dynamics. 2007;**12**(45):1097-1116

[4] Johansson A, Nielsen JCO. Out-of-round railway wheels-wheel rail contact forces and track response derived from field tests and numerical simulations. Proceedings of the Institution of Mechanical Engineers Part F: Journal of Rail & Rapid Transit. 2003;**217**(2):135-146

[5] Jing L, Han LL. Further study on the wheel–rail impact response induced by a single wheel flat: The coupling effect of strain rate and thermal stress. Vehicle System Dynamics. 2017. DOI: 10.1080/00423114.2017.1340651

[6] Steenbergen MJMM. Wheel-rail interaction at short-wave irregularities [Doctoral thesis]. Delft: Netherlands, Delft University of Technology; 2008

[7] Han LL, Jing L, Zhao LM. Finite element analysis of the wheel–rail impact behavior induced by a wheel flat for high-speed trains: The influence of strain rate. Proceedings of the Institution of Mechanical Engineers Part F: Journal of Rail & Rapid Transit. 2017. DOI: 10.1177/0954409717704790

[8] Wu TX, Thompson DJ. A hybrid model for the noise generation due to railway wheel flats. Journal of Sound & Vibration. 2002;**251**(1):115-139

[9] Uzzal RUA, Ahmed AKW, Bhat RB. Modelling, validation and analysis of a three-dimensional railway vehicle-track system model with linear and nonlinear track properties in the presence of wheel flats. Vehicle System Dynamics. 2013;**51**(11):1695-1721

[10] Pletz M, Daves W, Ossberger H. A wheel set/crossing model regarding impact, sliding and deformation—Explicit finite element approach. Wear. 2012;**294-295**:446-456

[11] Zhao X, Wen ZF, Zhu MH. A study on high-speed rolling contact between a wheel and a contaminated rail. Vehicle System Dynamics. 2014;**10**(52):1270-1287

[12] Polach O. Creep forces in simulations of traction vehicles running on adhesion limit. Wear. 2005;**258**:992-1000

[13] Torstensson PT, Nielsen JCO. Simulation of dynamic vehicle–track interaction on small radius curves. Vehicle System Dynamics. 2011;**49**:1711-1732

[14] BS-EN-13104 Standard. Railway applications-Wheelsets and bogies-Powered axles-Design method; 2009

[15] LS-DYNA Theory Manual. Livermore Software Technology Corporation; California, 2006

[16] Jing L, Su XY, Zhao LM. The dynamic compressive behaviour and constitutive modelling of D1 railway wheel steel over a wide range of strain rates and temperatures. Results in Physics. 2017;**7**:1452-1461

[17] Tian Y, Cheng YR, Liu XW. Studies on the dynamic behaviour of U71Mn rail steel under high strain rates. China Railway Science. 1992;**13**:34-42

10

3D Digital Simulation for Material Damage Mechanism Identification in a Railway Carriage Pressure Vessel

Alessandra Caggiano and Roberto Teti

Abstract

Digital simulation approaches applied to railway engineering allow to investigate different railway scenarios via 3D digital twins of real objects, motion simulation, and collision detection to identify the root causes of critical damage and estimate the most likely sources of railway accidents. In this work, a digital simulation approach is applied to a real catastrophic train accident in which a railway carriage carrying a pressure vessel collided with an obstacle that generated a cut in the pressure vessel casing. This cut initiated a liquefied petroleum gas leakage that expanded in the environment and caused the explosion and blaze responsible for human casualties. Traditional railway accident reconstruction procedures identified two potential objects accountable for the cutting of the pressure vessel casing: a wing rail and a track reference stake. Based on digital terrain models and reconstructed models of the railway carriage, 3D digital simulation scenarios were created to detect every possible collision of the pressure vessel with the infrastructure environment and investigate whether the shape of the cut in the pressure vessel wall fits the damage visible on the obstacles and whether the interference between obstacle and pressure vessel wall could generate the chip through an interaction similar to metal cutting.

Keywords: railway carriage, pressure vessel, safety, material damage, catastrophic accident, digital twin

1. Introduction

The employment of digital simulation approaches in railway engineering enables to explore various railway scenarios through the use of 3D digital twins of real objects, motion simulation, and collision detection, which represent valuable tools to detect the original causes of critical damage and estimate the most liable sources of railway accidents.

In a real catastrophic train accident, a railway carriage carrying a pressure vessel collided with an obstacle that generated a cut in the casing of the pressure vessel, initiating a liquefied petroleum gas leakage that expanded in the environment and caused the explosion and fire responsible for a high number of human casualties.

The utilization of traditional railway accident reconstruction procedures allowed to recognize two objects potentially accountable for the cutting of the pressure vessel casing: a wing rail and a track reference stake. In this work, based on digital terrain models and reconstructed models of the railway carriage obtained via reverse engineering, different 3D digital simulation scenarios are created, allowing to move and rotate the carriage interactively, as well as to examine the accident scene from different points of view, with the aim to detect every possible collision of the pressure vessel with the infrastructure environment. The scope of the 3D digital simulation employment is to investigate whether the form of the cut in the pressure vessel wall fits to the damage visible on the two potential objects accountable for the cutting of the pressure vessel casing and whether the interference between object and pressure vessel wall could generate the chip through an interaction similar to metal cutting.

2. Catastrophic train accident description and reconstruction

On the 29th of June 2009, around midnight, a catastrophic train accident occurred in the train station of Viareggio, Italy [1]. The train was a freight train with 14 carriages carrying pressure vessels filled with liquefied petroleum gas. After the accident, the whole train divided up into five separated parts: part 1 was the locomotive, still upright on the rails; part 2 was carriage no. 1, reversed on its left side; part 3 consisted of 2 carriages, nos. 2 and 3, still connected to each other and reversed on their left side; part 4 consisted of 2 carriages, nos. 4 and 5, still connected to one another and reversed on their left side; and part 5 was subdivided into two halves: the first half of part 5 consisted of five carriages, from carriage no. 6 to carriage no. 10, connected to each other and derailed but not reversed. The second half of part 5 consisted of four carriages, from carriage no. 11 to carriage no. 14, connected to each other and upright on the rails.

Each carriage carrying a pressure vessel had a front bogie and a back bogie; each bogie had a first (front) axle and a second (rear) axle [2–4].

The train derailment was caused by the sudden failure of the second axle of the front bogie of carriage number 1 [5, 6]. The axle breakage was caused by rotating flexural fatigue, determining a fatigue crack extension up to more than 65% of the axle resisting section. **Figure 1** [6] shows that the axle was fractured externally with respect to the left wheel (**Figure 1a**), and this failure was the original cause of the train derailment. **Figure 1b** illustrates the characteristic fatigue failure of the fracture surface of the axle on the wheel side: presence of a smooth area related to the slow propagation of the fatigue crack (upper zone in the image) and of a rough area of rapid failure of the residual axle section (lower zone in the image). The same characteristic fatigue failure is observable on the matching axle fracture surface on the axle box side.

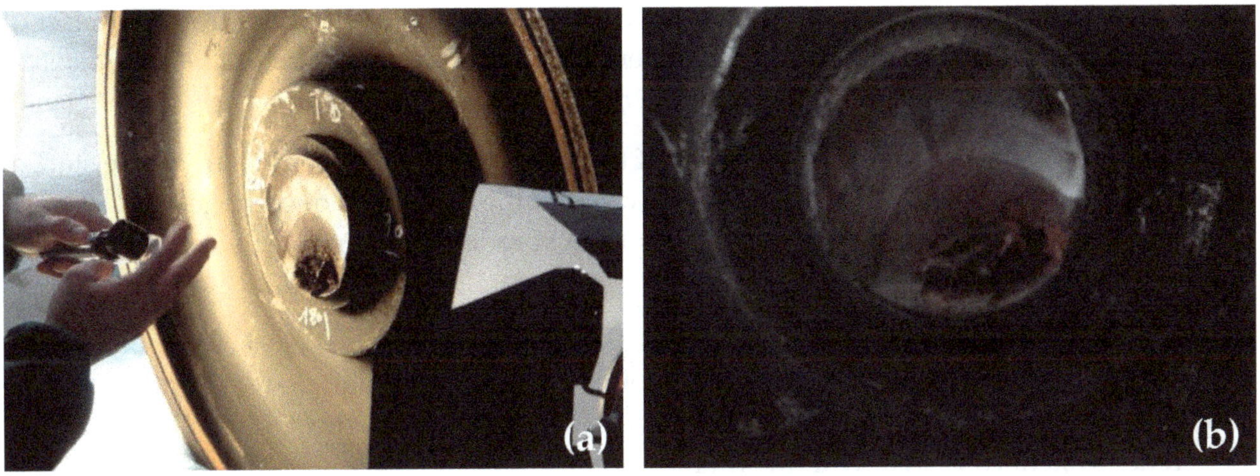

Figure 1. (a) Fatigue failure of the fracture surface of the axle on the wheel side. (b) Presence of a smooth area related to the slow propagation of the fatigue crack and of a rougher area of rapid failure of the residual axle section.

As regards the right wheel of the second wheelset (rear) of the front bogie of carriage no. 1, the right extremity of the axle did not break, but clear signs of corrosion are evidenced at the fillet between the axle extremity, called axel spindle, with a smaller diameter, and the axle zone with a larger diameter for shrink fitting of the wheel.

Figure 2 shows the bogie model Y25 mounted on the pressure vessel carriages composing train no. 50325, with indication of some of its components, including the axle boxes. Due to a fatigue crack in its final development phase transversally with respect to the wheelset axle, the vertical rigidity of the axle box supported by the axel spindle was notably reduced up to reaching a nonequilibrium condition of the vertical forces pressing the wheels on the rails. The surmounting of the rail occurred due to a notable reduction of the vertical force acting on the right front wheel, diagonally opposed to the force acting on the weakened axle spindle (left rear wheel).

Under these conditions, the transversal (horizontal) force acting on the right front wheel, which was no longer loaded downward by a sufficient vertical force, was high enough to

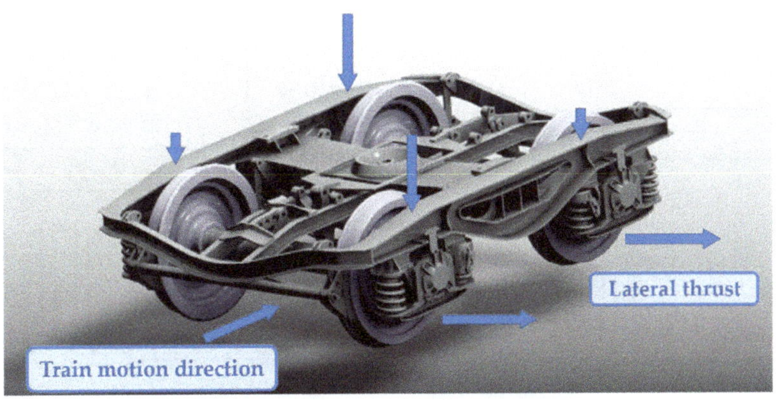

Figure 2. Forces acting on the first bogie of carriage no. 1 after breakage of the second axle (rear wheelset) due to fatigue failure externally with respect to the left wheel.

make the wheel climb over the top of the rail, causing the derailment of the front wheelset of the first bogie of carriage no. 1. The derailment start was located at km 120 + 265, next to Viareggio station entry when coming from the north (from Genoa station).

Figure 3 shows the traces of the initial phase of the derailment and in particular: (a) the start of the surmounting of the rail by the right front wheel of the first bogie of carriage no. 1. (b) The sleepers cut by the left wheel and the sidewalk curb abraded by the right front axle box of the first bogie of carriage no. 1.

Figure 4 shows the relative positions, after the derailment, of carriage no. 1, the rotated front bogie, the right front axle box abrading the sidewalk curb, and the abraded sidewalk curb. The derailed bogie rotated around its axis in clockwise direction with respect to the vertical axis in a reference system consistent with the train.

Near the end of the sidewalk, in the southward direction (toward Pisa station), there is a walkway at grade endowed with a slanted ramp (**Figure 5**). Until the walkway at grade, the derailed carriage had been guided by the sidewalk curb that acted on the highly abraded right front axle box just like a grinding wheel on a workpiece in a grinding process.

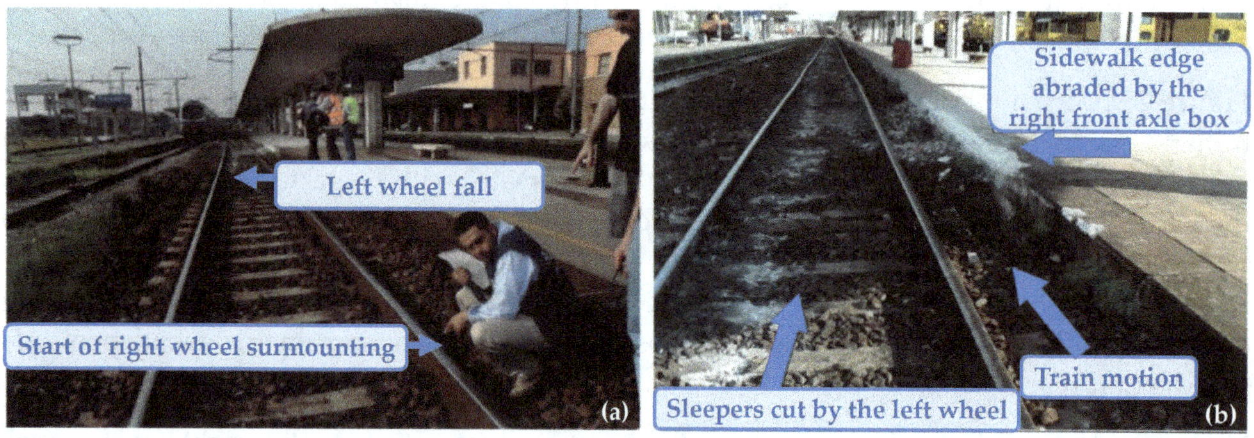

Figure 3. Traces of the initial phase of the derailment. (a) Start of the surmounting of the rail by the right front wheel of the first bogie of carriage no. 1. (b) Sleepers cut by the left wheel and sidewalk edge abraded by the right front axle box of the first bogie of carriage no. 1.

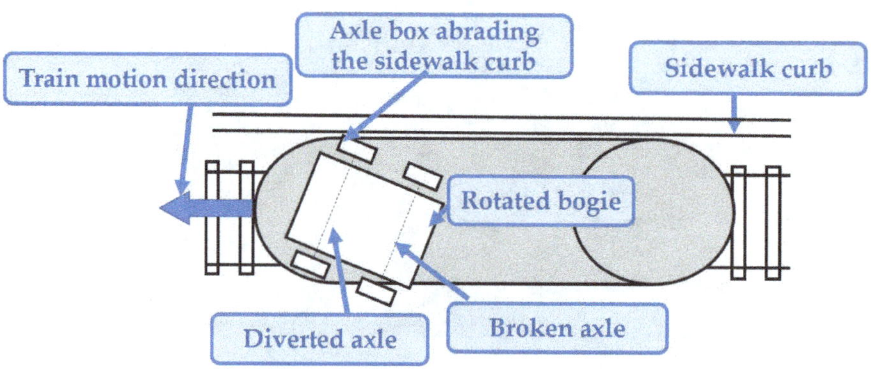

Figure 4. Relative positions, after derailment, of carriage no. 1, right front axle box, and sidewalk curb.

Figure 5. Walkway at grade endowed with a slanted ramp on the sidewalk, near the track of the derailed train, and trace on the granite curb left by the flange of the right wheel of the front wheelset of the first bogie of carriage no. 1.

Upon arriving at the slanted ramp of the walkway at grade, the front bogie of carriage no. 1, deprived of the guide provided on its right by the curb of the sidewalk and still pulled by the locomotive, engaged the slanted ramp with the right wheel of the first wheelset, climbing on the sidewalk and leaving on it the trace of the wheel flange as shown in **Figure 5**.

The upward movement of the carriage due to the climbing provoked its overturning on its left side in the train motion direction, as described below.

Figure 6. Concrete sidewalk damaged by the right wheel of the front wheelset of the first bogie of carriage no. 1 and starting point of the carriage overturning.

Figure 7. Superstructure zone wrecked due to the "plowing" mechanism of the track by the front wheelset of the first bogie.

Few meters after the end of the walkway at grade, the curb material, made of granite up to that point, changed into a concrete structure that completely crumbled under the action of the wheel due to much lower strength of the concrete material (see **Figure 6**).

At the end of the concrete sidewalk zone, crumbled by the right wheel of the front wheelset of the first bogie, the overturning of carriage no. 1 on its left side in the train motion direction started. The carriage continued its course till it landed at about 73 m from the point where its detachment from the track began. The landing point of carriage no. 1 corresponds to the

Figure 8. Plowing mechanism of the track by the bogie of the overturned carriage.

beginning of the superstructure zone wrecked by the "plowing" mechanism of the track, as shown in **Figure 7**.

When, at the end of the overturning phase, carriage no. 1 crashed onto the infrastructure, it presented the remaining wheelset still mounted on its first bogie with its axis slightly slanted forward and the wheel, originally on the left side in the train motion direction, stuck under the sleepers; as the carriage proceeded its course under these conditions, the track was plowed according to the mechanism illustrated in **Figure 8**.

After the end of the plowing zone, the carriage continued to slide on its left side along the railway superstructure.

3. Cut on the tank shell of carriage no. 1 and elements potentially responsible for the opening of the cut

In the final phase of the accident, shortly before carriage no. 1 finally ended its course, an external body present on the railway superstructure hit the pressure vessel and opened the cut on the tank shell shown in **Figure 9**.

From this cut, the liquefied petroleum gas came out in liquid phase under pressure and started the liquid-gas phase transformation, creating a gaseous cloud that expanded in the environment. After about 190 s, accidental causes determined the ignition of the gaseous cloud and the catastrophic consequences of the accident.

Figure 10 shows the aerial view of the accident scenario with indication of the potentially responsible elements that could have generated the cut on the tank shell through a mechanism similar to the action of a metal cutting tool. The positions of the elements are reported in more details in **Figure 11**.

- Track reference stake no. 23, found under the pressure vessel at about 7 m from the back bumper of carriage no. 1.

Figure 9. Carriage no. 1 with a cut opening on the tank shell.

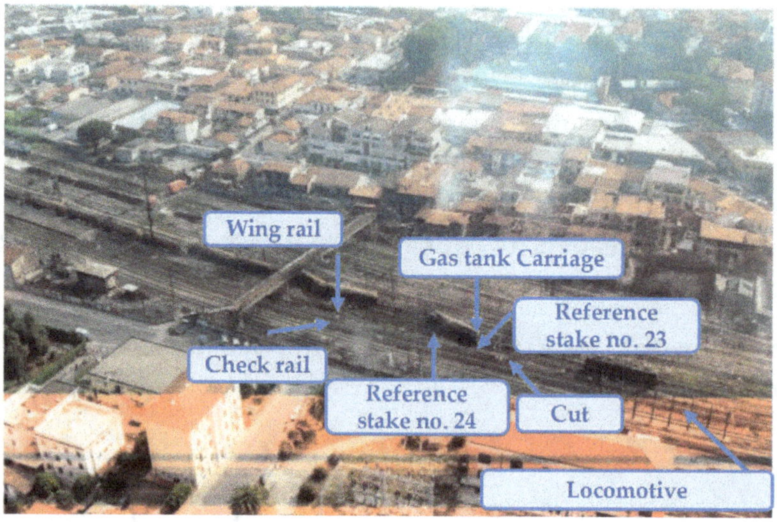

Figure 10. Aerial view of the accident with indication of the elements potentially responsible for the opening of the cut on the tank shell.

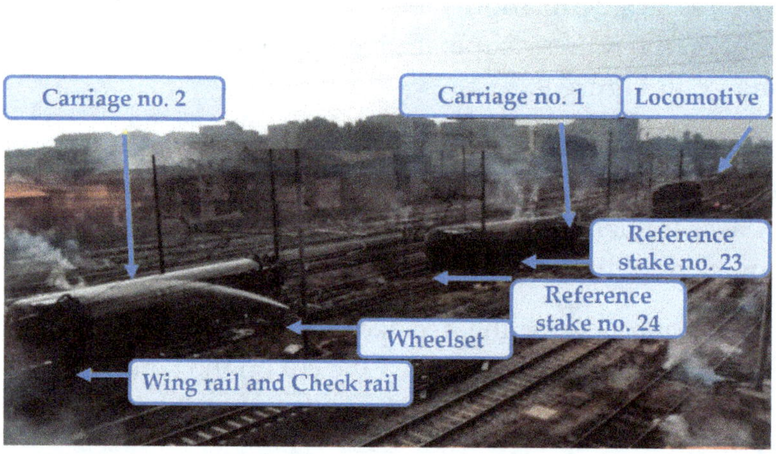

Figure 11. Positions of the elements potentially responsible for the cut generation on the tank shell through an action similar to metal cutting: track reference stake nos. 23 and 24, wing rail and check rail.

- Track reference stake no. 24, found at about 3 m before the back bumper of carriage no. 1.
- Wing rail and check rail [2–4] that may have been hit by diverse components of the rolling stock.

3.1. Track reference stake nos. 23 and 24

It is worth mentioning that a track curve reference stake is made of a chunk of rail anchored in a concrete plinth grounded in the ballast so that the reference stake protrudes vertically out of the ground level with an elevation 5 cm higher than the highest rail of the track curve. The function of the track reference stakes is to allow for the periodical checking of the correct geometry of the track curve next to which they are installed. To perform this checking, a technician periodically controls that the internal edge of the highest rail of the

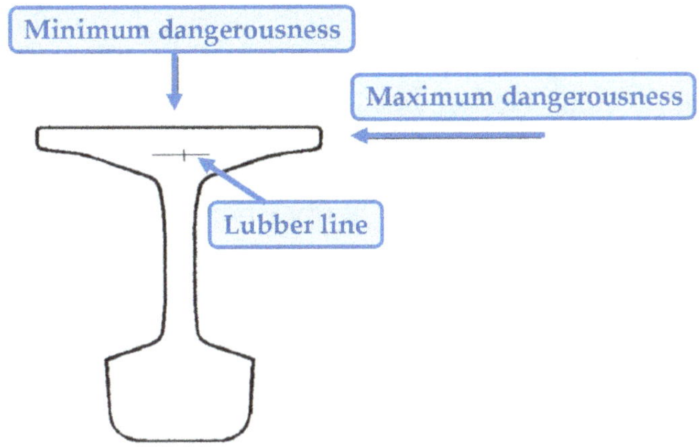

Figure 12. Transverse section of a track curve reference stake and directions of minimum and maximum danger grade in case of impact by a moving body against the track reference stake.

track curve is 1.5 m distant from the reference line traced on the top section of the track reference stake.

In **Figure 12**, the transverse section of a track curve reference stake is shown illustrating the variation of the degree of danger as a function of the impact direction of a moving body against the track reference stake. In the figure, two possible impact directions are considered: the one of maximum degree of danger, parallel to the lower surface of the rail chunk and the other of minimum degree of danger, perpendicular to the lower surface of the rail chunk. For intermediate impact angles, the danger increases going from a perpendicular impact to a parallel impact with reference to the lower surface of the rail chunk. In the latter case, the extremely high contact forces due to the sharp configuration of the corner of the lower surface of the rail chunk can easily determine the cutting of the sheet metal of the impacting tank shell.

Figure 13. Rear view of carriage no. 1 in the accident scene and positions of track reference stake nos. 23 and 24.

Figure 13 shows the rear view of carriage no. 1 in the accident scene and the positions of track reference stake nos. 23 and 24; the two track reference stakes are 10 m apart, as set by the technical regulations of the Italian railways.

Figure 14 shows the front view of carriage no. 1 where it can be seen that the front bumper lies on the right rail in the train motion direction of track no. 4.

In **Figure 15**, a detail of track reference stake no. 24, found under the tank of carriage no. 1 almost completely buried in the track ballast, is shown. The sharp corner of the lower surface of the rail chunk, making up the track reference stake, appears highly abraded with its abraded metal surface still not subjected to oxidization. This confirms that the abrasion process, due to sliding against the carriage, was very recent during the examinations carried out in the accident scenario.

Figure 14. Front view of carriage no. 1 in the accident scene and position of its front bumper.

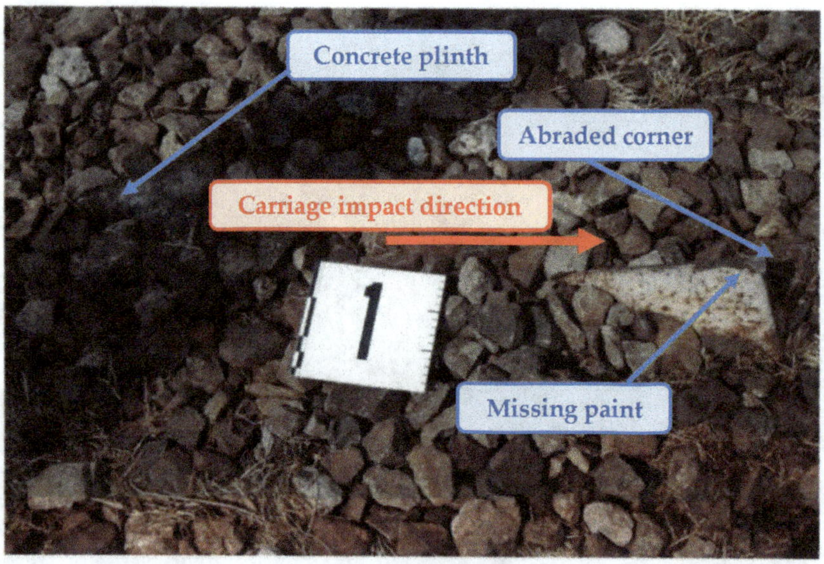

Figure 15. Track reference stake no. 24, photographed during the investigations on the accident site conditions, almost completely buried in the track ballast following its impact against the pressure vessel of carriage no. 1.

Figure 16 shows track reference stake no. 24 after its extraction from the ballast: there are no traces of flexural deformation evidencing that the track reference stake had been solely rotated in a rigid way on its plinth during its knocking down under the impact action of the pressure vessel of carriage no. 1. Moreover, the figure shows the notable extension of the abrasion of the sharp corner of the lower surface of the track reference stake [6].

In **Figure 17**, a detail of track reference stake no. 23 knocked down by the pressure vessel of carriage no. 1 about 10 m after track reference stake no. 24 had been hit is shown. The track reference stake no. 23 was found under the tank, and its sharp corners did not show significant traces of abrasion. At the moment of impact against track reference stake no. 23, the carriage was by then almost steady and therefore endowed with a much lower kinetic energy with respect to the moment of impact against track reference stake no. 24.

3.2. Wing rail and check rail

Figure 18 shows a detail of the deformation of the wing rail. The impact on the wing rail has a component in the transverse direction, oriented toward the left side in the train motion direction, and there are no traces of abrasion due to sliding on the deformed zone of the wing rail.

Figure 16. Track reference stake no. 24 photographed after its extraction from the track ballast. (a) In the foreground: flange of the track reference stake with damaged edge; in the background: view of the concrete plinth; and (b) and (c) details of the damaged edge.

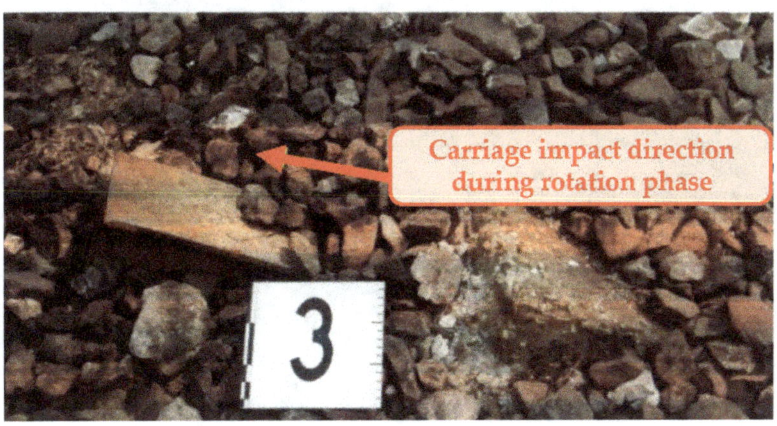

Figure 17. Track reference stake no. 23, photographed during the investigations on the accident site conditions, partially buried in the track ballast following its impact against the pressure vessel of carriage no. 1.

Figure 18. Wing rail photographed during the investigations at the accident site conditions.

Figure 19. Check rail opposite to the wing rail photographed during the investigations on the accident site conditions.

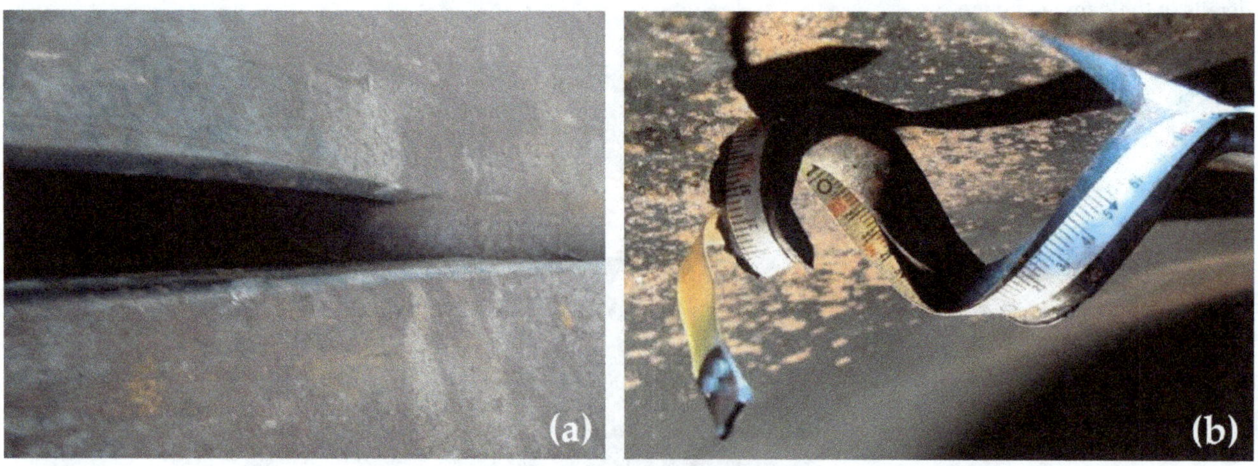

Figure 20. (a) Enlarged external view of the end side of the cut with the attached corresponding chip bent internally inside the tank shell and (b) enlarged internal view of the chip portion still attached to the tank shell.

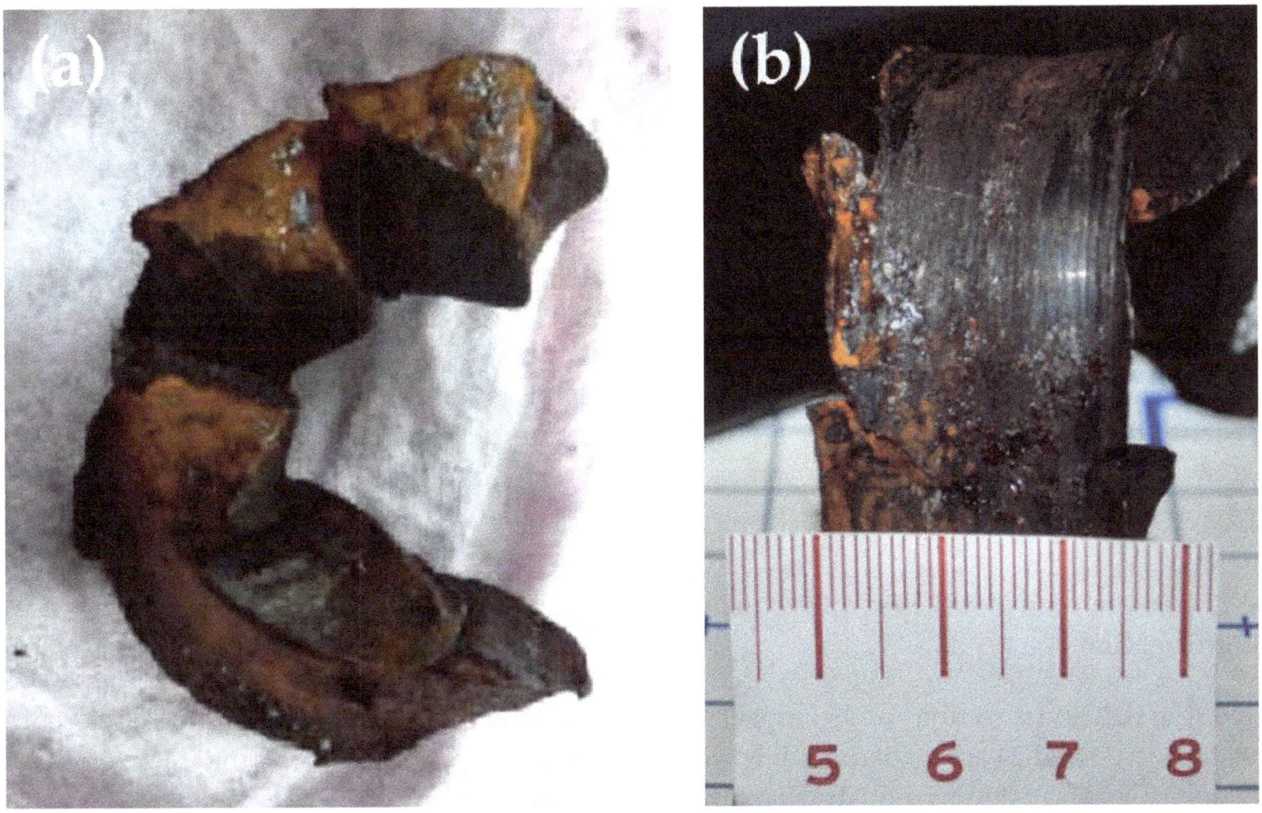

Figure 21. Chip portion detached from the tank shell showing the segmented morphology typical of a chip generated during a machining operation.

Figure 19 shows a detail of the deformation of the check rail opposite to the wing rail. It has been subject to a strong impact in the direction of motion of the overturned carriage no. 1. One element of the derailed material, sliding along the track, has hooked the check rail determining on it a cut and some bulges.

The final part of the chip remained attached to the tank shell (**Figure 20**), while the initial chip portion was detached from the shell and was found inside the tank. This segmented chip shows the typical morphology of a chip formed during a machining operation that generates a notable chip segmentation (**Figure 21**).

4. 3D Digital modeling and simulation analysis to identify the element responsible for the cut

To identify the element that generated the cut in the tank shell of carriage no. 1 through a mechanism similar to the action of a metal cutting tool, an approach based on 3D digital modeling and simulation was employed [7].

Digital simulation allows to investigate different scenarios through the use of 3D digital twins of real objects, motion simulation and detection of interferences, which are useful to identify the root causes of the critical cut damage on the tank shell.

Within the digital simulation scenarios, it was investigated whether the form of the cut in the pressure vessel fits to the damage found on the elements potentially accountable for the cutting of the tank shell and whether the interference between elements and tank shell could generate the chip through an interaction similar to metal cutting.

The 3D digital twins of the railway carriage and the elements potentially accountable for the cutting were obtained by acquiring the geometry and shape of the real parts and reconstructing their digital models via a reverse engineering (RE) procedure [8–10].

RE is the process of acquiring the geometry and shape of a part and reconstructing its digital model. The first stage of the RE procedure, i.e., digital data acquisition, can be carried out by means of several different tools [9, 11]. In this case, a time-of-flight laser sensor and a portable laser scanner with high accuracy were employed for scanning the railway carriage and the surrounding elements.

4.1. Track reference stake no. 23 exclusion and track reference stake no. 24 initial confirmation

In **Figure 22**, the directions of the impact between the tank of carriage no. 1 and track reference stake nos. 23 and 24 are evidenced. It is worth recalling that the tank first hit the track reference stake no. 24 and afterward the track reference stake no. 23.

During the interaction with track reference stake no. 24, which was hit first, the tank, under the action of the applied forces, initiated a rotational motion around a vertical axis, as endorsed by the arched shape of the cut on the tank shell.

Figure 23 illustrates the motion of tank carriage no. 1 when, after hitting track reference stake no. 24, it proceeded to hitting track reference stake no. 23. The rotation of the carriage around the vertical axis was about 5° clockwise for an observer looking at the scene from above.

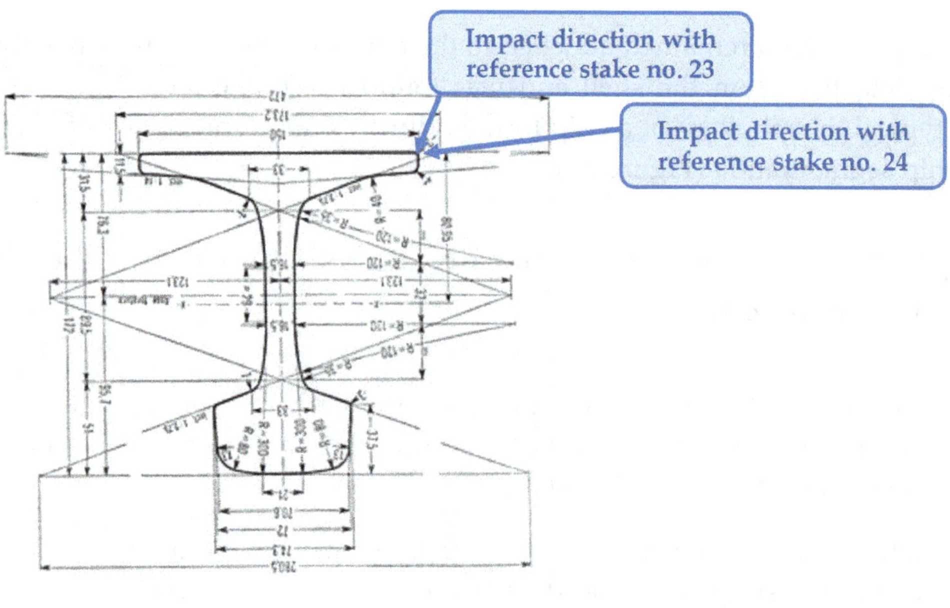

Figure 22. Direction of the impact between the tank of carriage no. 1 and track reference stake nos. 23 and 24.

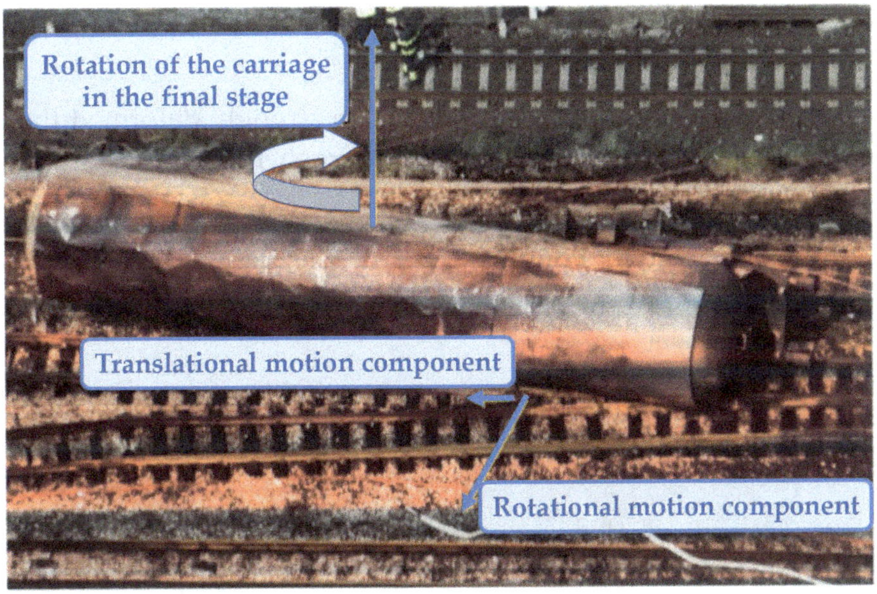

Figure 23. Rotational motion of tank carriage no. 1 around a vertical axis.

4.2. Check rail exclusion

Figure 24 shows a series of details of the core zone of railway switch 13B. The scene refers to the zone around the wing rail and the check rail as it appeared after the accident, where the following elements were found:

- The anchorage bracket of carriage no. 1, sitting at about 3.5 m beyond the position of the wing rail.
- A chunk of the right-side axle box, in the train motion direction, of the first (front) wheel set no. 85890 of the front bogie, i.e., the axle box which remained mounted on carriage no. 1 after the second (back) wheel set no. 98331 of the front bogie was detached.
- Part of the sheet metal billboard welded on the side of carriage no. 1 for the application of transport documentation.

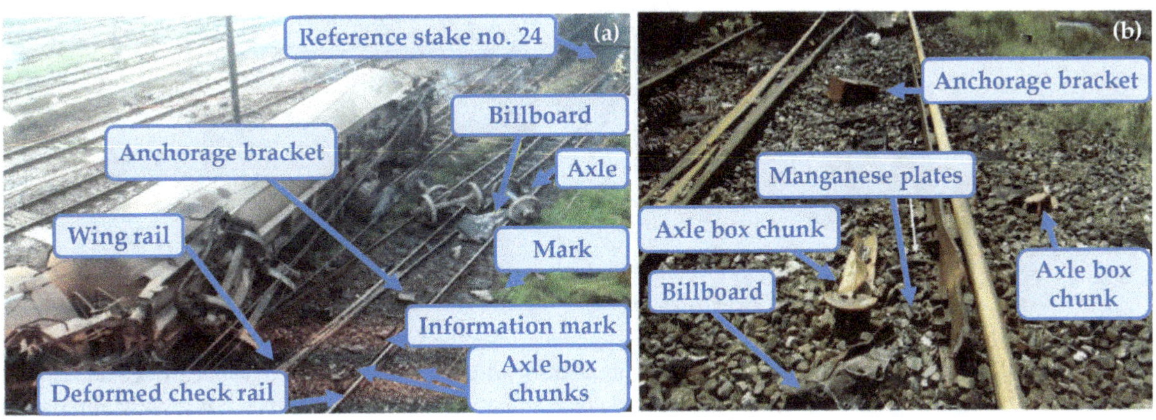

Figure 24. Zone of the heart of railway switch 13B around the wing rail and the check rail immediately after the accident.

- The manganese plates making up the kinematic surface on which the axle boxes slide with respect to the axel guard

All these wreckages demonstrate that the body, which strongly deformed the check rail, was an axle box that was practically destroyed during the impact (**Figure 25**).

Figure 26 shows two of the three chunks of the right axle box of the first wheelset of the front bogie of carriage no. 1, whereas **Figure 27** shows the digital model of the fragment of **Figure 26b** obtained by reverse engineering reconstruction based on geometry acquisition with a structured light optical sensor.

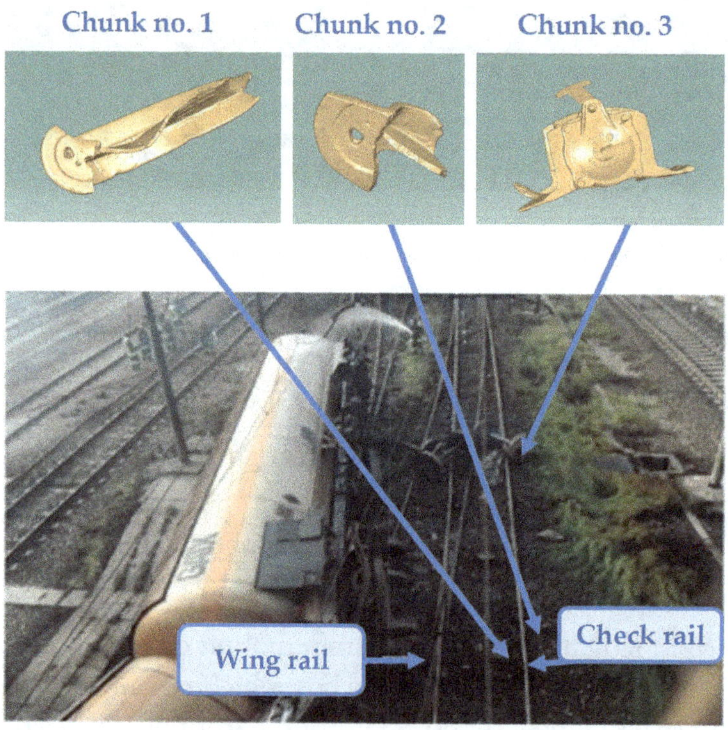

Figure 25. Position, in the accident scene, of the chunks of the right axle box of the front wheelset of the front boogie of carriage no. 1; the axle box was greatly abraded and broken into three fragments.

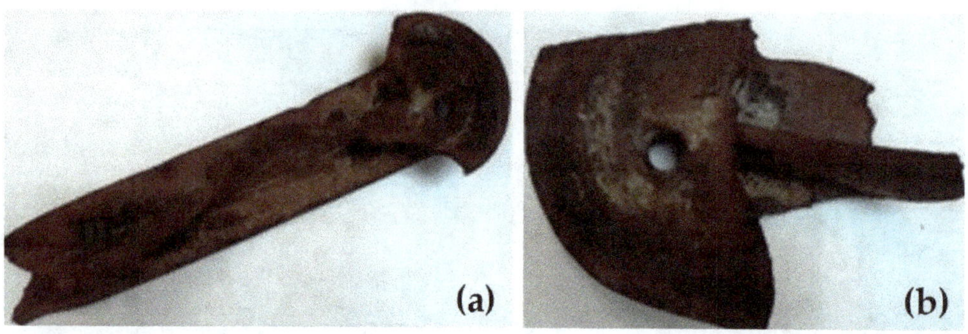

Figure 26. Two of the three chunks of the right axle box of the front wheel set of the front bogie of carriage no. 1: (a) chunk of the axle box found between the rails near the deformed check rail and (b) chunk of the axle box greatly abraded by the granite sidewalk curb exerting a grinding action on the right axle box, found externally to the track near the right-side rail in the train motion direction.

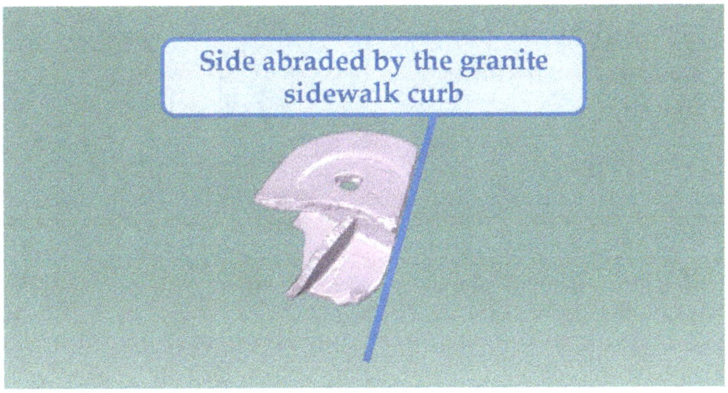

Figure 27. Digital twin of the chunk of the right axel box of the first wheel set of the front bogey of carriage no. 1 obtained by reverse engineering reconstruction based on geometry acquisition with a structured light optical sensor.

The front part of the right axle box—right side in the train motion direction—of the front wheel set was greatly abraded by the granite of the sidewalk curb that acted on the right front axle box just like a grinding wheel on the workpiece in a grinding process (**Figures 26b** and **27**).

The realization of the digital twins of the check rail and the right axle box of the first wheel set of the front bogie of carriage no. 1, obtained by reverse engineering reconstruction based on geometry acquisition with a structured light optical sensor, allowed to reconstruct under simulation the kinematics of the impact between axle box and check rail, as illustrated in **Figure 28a**. **Figure 28b** shows the traces of damage on the check rail, which confirm the kinematics illustrated in **Figure 28a**.

It is, therefore, evident that the damage on the check rail cannot be related to the opening of the cut on the tank shell of carriage no. 1.

4.3. Initial confirmation of both wing rail and track reference stake no. 24

Figure 29 shows the digital model of the tank of carriage no. 1, obtained by reverse engineering reconstruction based on geometry acquisition with a time-of-flight laser sensor. The digital model highlights the incidence angle of the cut that varies during progressing of the cutting process. The incidence angle is the angle between the generatrix of the cylindrical body of the tank and the straight-line tangent to the outline of the cut at the point of intersection between generatrix and cut.

Figure 30 shows the photograph of the cut present on the tank shell of carriage no. 1 (**Figure 30a**) and the corresponding digital model obtained by reverse engineering reconstruction based on geometry acquisition with a time-of-flight laser sensor (**Figure 30b**).

By examining the digital models of **Figures 29** and **30b**, it can be noted that the cut starts with an incidence angle of 5.1° and ends with an incidence angle of 9.8°. Accordingly, the rotation of the tank around a vertical axis, as illustrated in **Figure 12**, was about 5°.

In **Figure 31**, the digital model of the wing rail, obtained by reverse engineering reconstruction based on geometry acquisition with a structured light optical sensor, is reported in side view (**Figure 31a**) and longitudinal view (**Figure 31b**). It can be noted that the blunted surface of

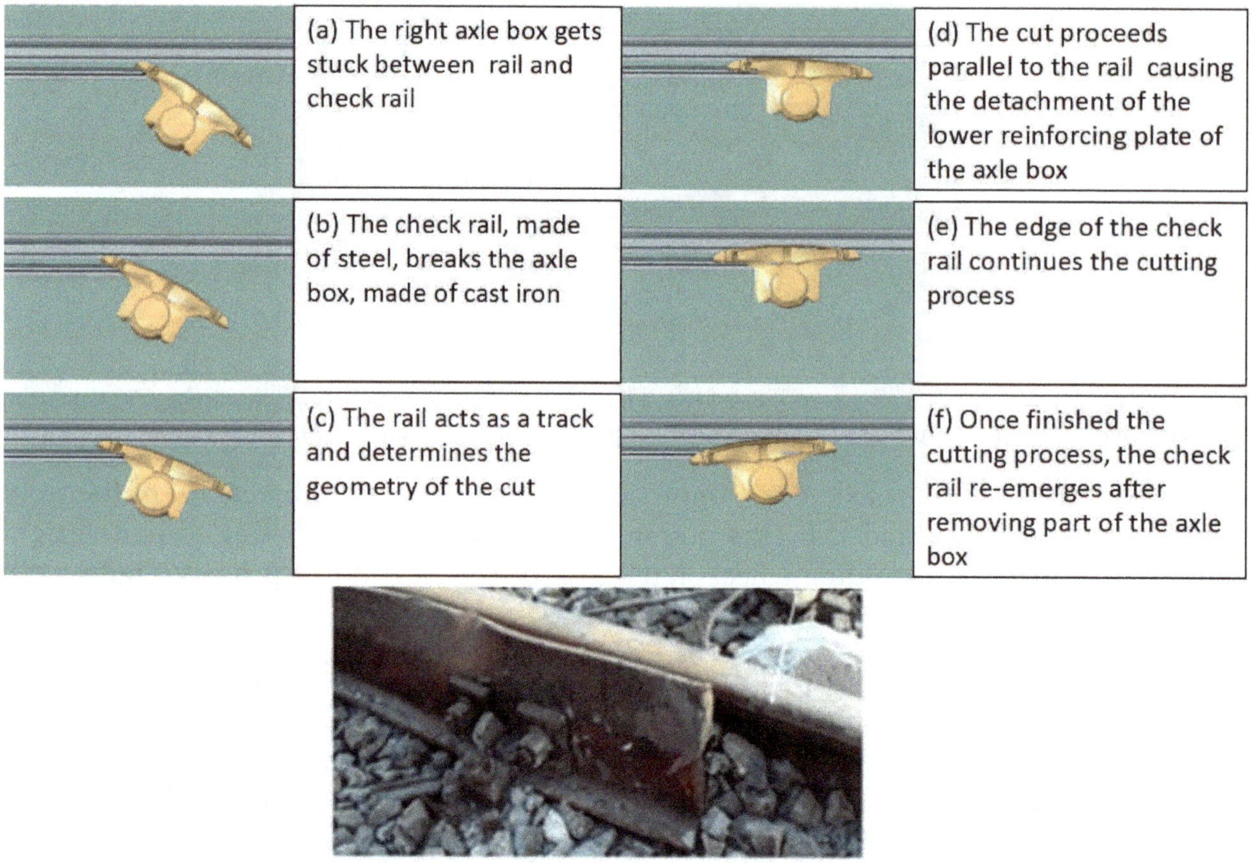

Figure 28. (a) Simulation reconstruction of the kinematics of the impact between check rail and right axel box of the first wheel set of the front bogey of carriage no. 1 and (b) traces of damage on the check rail confirming the kinematics described in **Figure 28a**.

the extremity of the wing rail interested by the damage does not evidence traces of abrasion due to sliding.

Figure 32 illustrates the simulation of the interaction between wing rail (the digital model of which was considered fixed with respect to the ground) and the deformed area of the tank

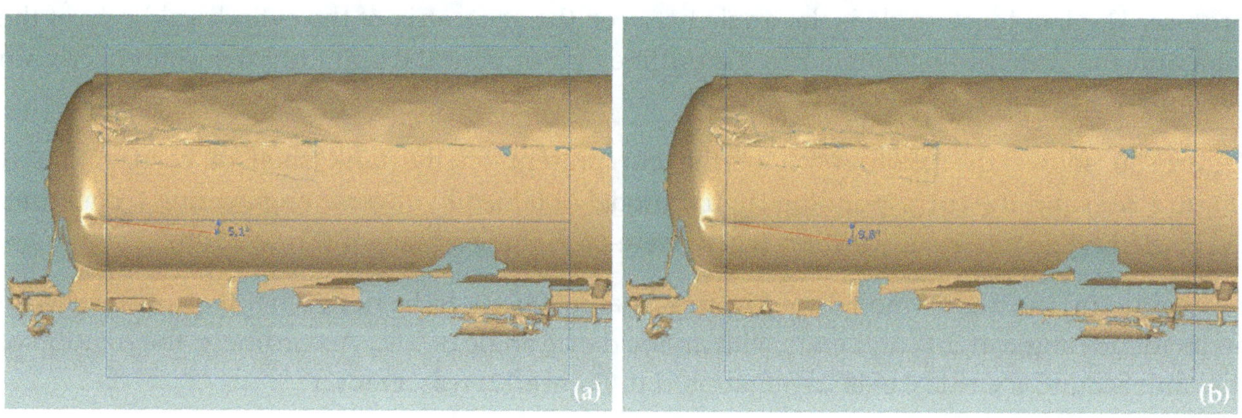

Figure 29. Digital twin of the tank of carriage no. 1. Obtained by reverse engineering reconstruction based on geometry acquisition with a time-of-flight laser sensor, with indication of the incidence angle of (a) beginning of the cut and (b) end of the cut.

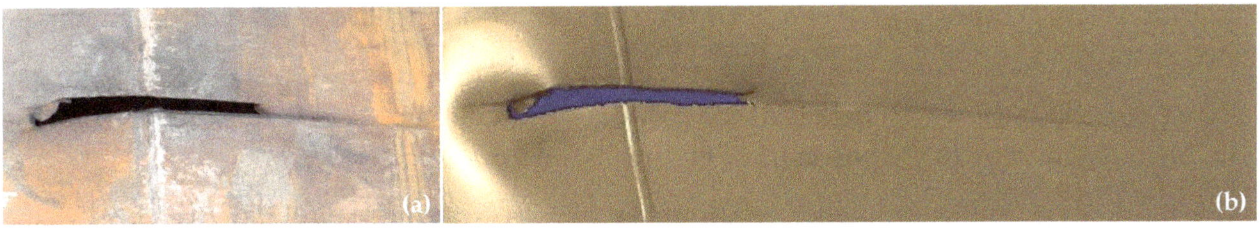

Figure 30. (a) Photograph of the cut present on the tank shell of carriage no. 1 and (b) digital model of the cut obtained by reverse engineering reconstruction based on geometry acquisition with a time-of-flight laser sensor.

shell of carriage no. 1. The simulation comprises the following phases: (a) initial contact between wing rail and shell of the tank, (b) penetration into the sheet metal of the tank shell and start of the cutting, (c) progressing of the cutting process, (d) end of the cutting, (e) start of the sliding on the tank shell surface, and (f) end of the sliding.

Figure 33a and **b** shows the details of the cutting phases corresponding to the positions of **Figure 32c** and **d**, respectively.

The analysis of the diverse relative positions of the wing rail and the cut opening shows a geometrical compatibility between the two digital models during the cutting process.

The same simulation was carried out by considering the damage caused by track reference stake no. 24, the digital model of which, realized by reverse engineering reconstruction based on geometry acquisition with a structured light optical sensor, is reported in **Figure 34**. The figure shows a view from the head side (**Figure 34a**) and from the flange side (**Figure 34b**) of the rail chunk making up the track reference stake, evidencing the highly abraded sharp corner.

Figure 35 shows the simulation of the interaction between track reference stake no. 24 and side of the tank shell of carriage no. 1, comprising the following phases: (a) initial contact between track reference stake and tank shell side, (b) penetration into the tank shell

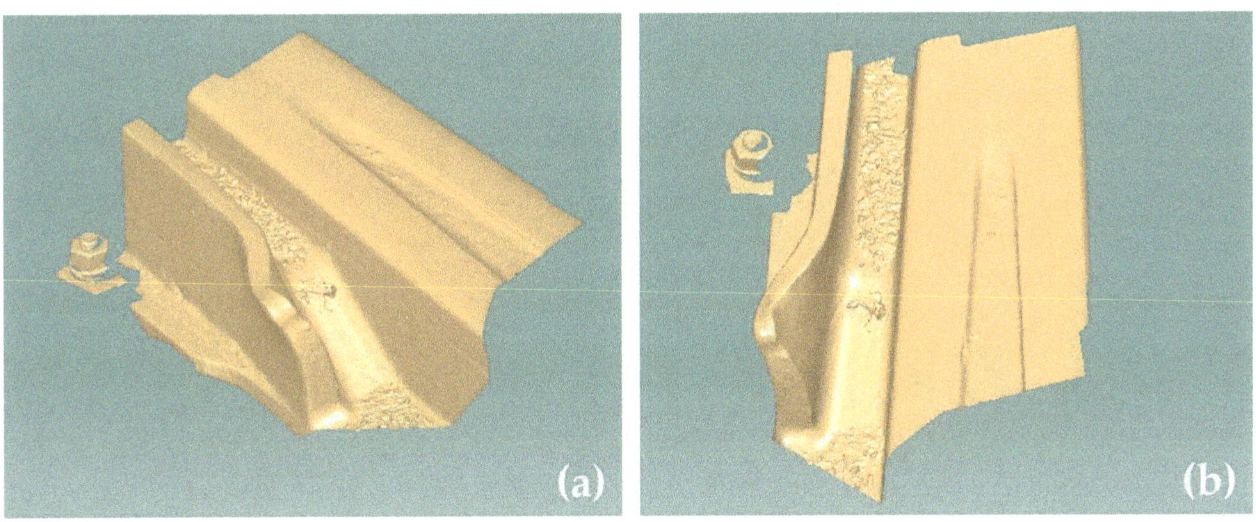

Figure 31. Digital twin of the wing rail obtained by reverse engineering reconstruction based on geometry acquisition with a structured light optical sensor: (a) side view and (b) longitudinal view.

side and start of cutting (c), (d) end of cutting, (e) start of sliding, and (f) end of sliding on the shell surface.

Figure 36a and **b** shows the details of the phases of the cutting action by track reference stake no. 24 corresponding to the positions of **Figure 35c** and **d**, respectively.

Also the analysis of this simulation confirms the geometrical compatibility between the shape of track reference stake no. 24 and the opening generated by its cutting action on the tank shell,

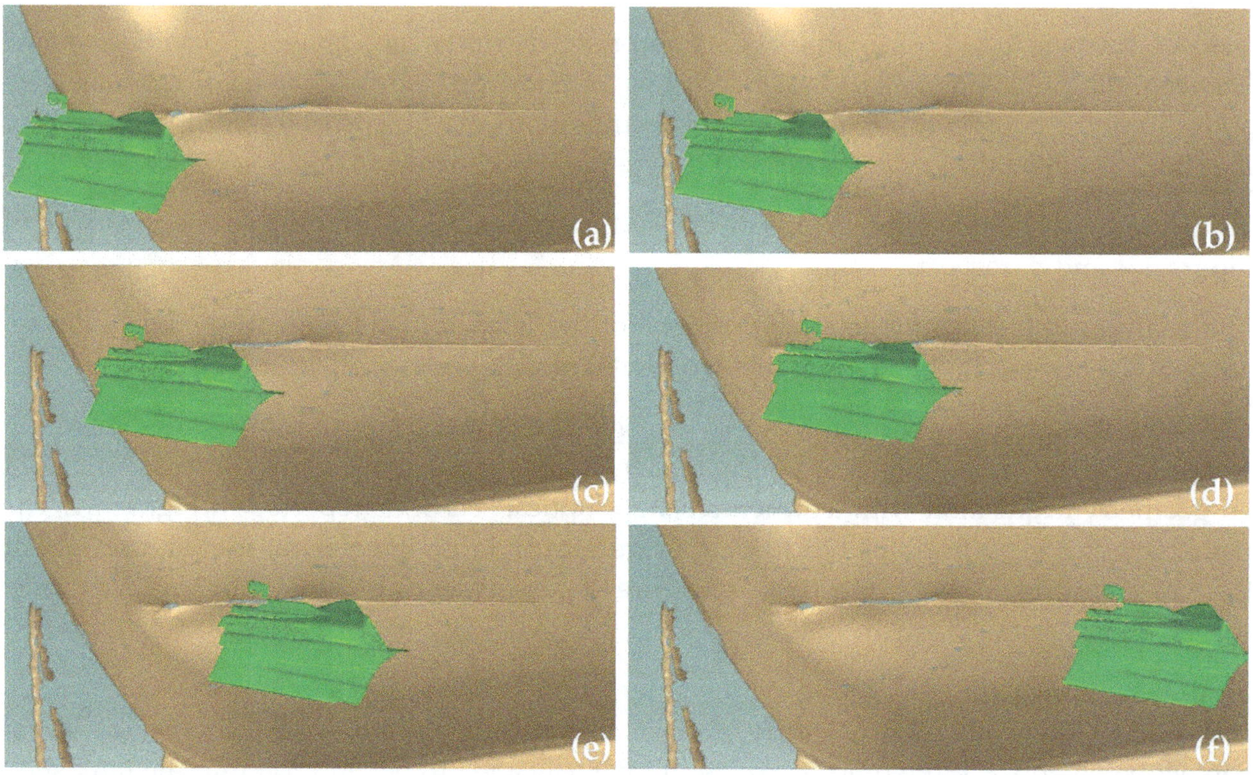

Figure 32. Simulation of the interaction between wing rail and tank shell of carriage no. 1: (a) contact, (b) penetration, (c) cutting, (d) exit, (e) start of sliding, and (f) end of sliding.

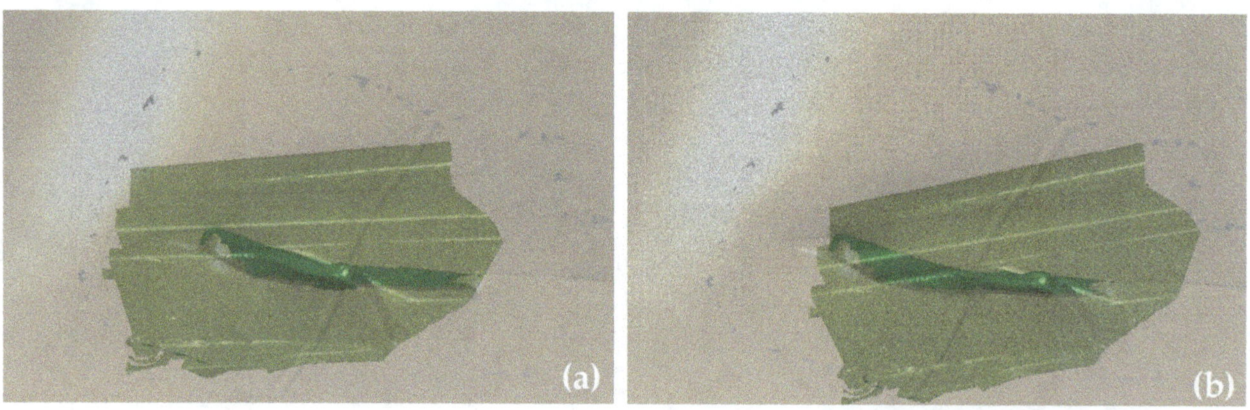

Figure 33. Details of the phases of the cutting action by the wing rail corresponding to: (a) the position of **Figure 32c** and (b) the position of **Figure 32d**.

taking into account the yielding and knocking down of track reference stake no. 24 under the impact of tank carriage no. 1.

Thus, the analyses of the damage generation process, carried out by simulating the relative displacements between the deformed zone of the tank shell and the possible responsible elements encountered by carriage no. 1 during the accident, evidenced that both wing rail

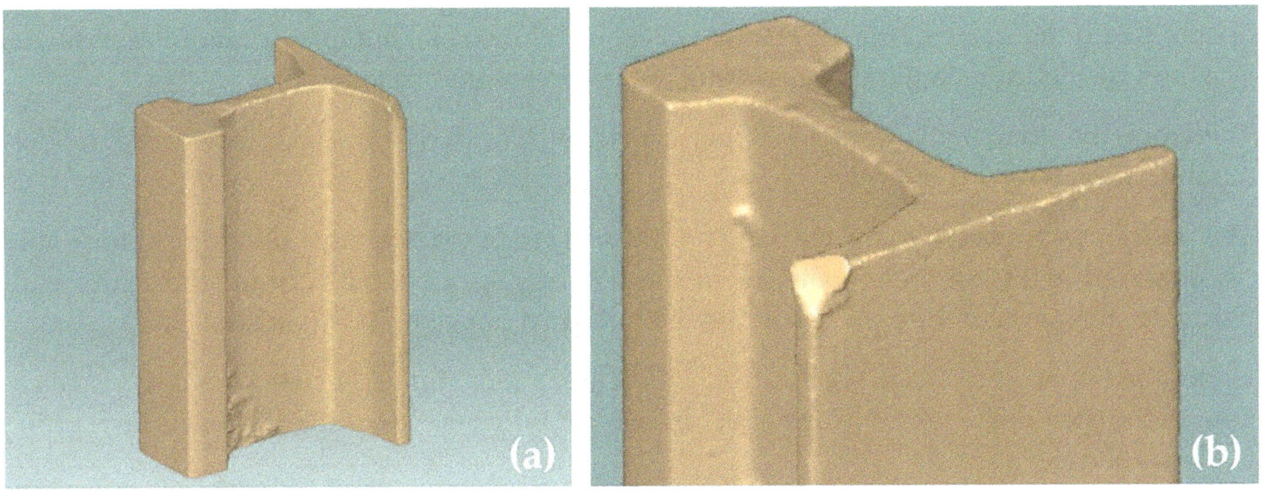

Figure 34. Digital twin of track reference stake no. 24, obtained by reverse engineering reconstruction based on geometry acquisition with a time-of-flight laser sensor: (a) view from the head side and (b) view from the flange side, evidencing the highly abraded sharp corner.

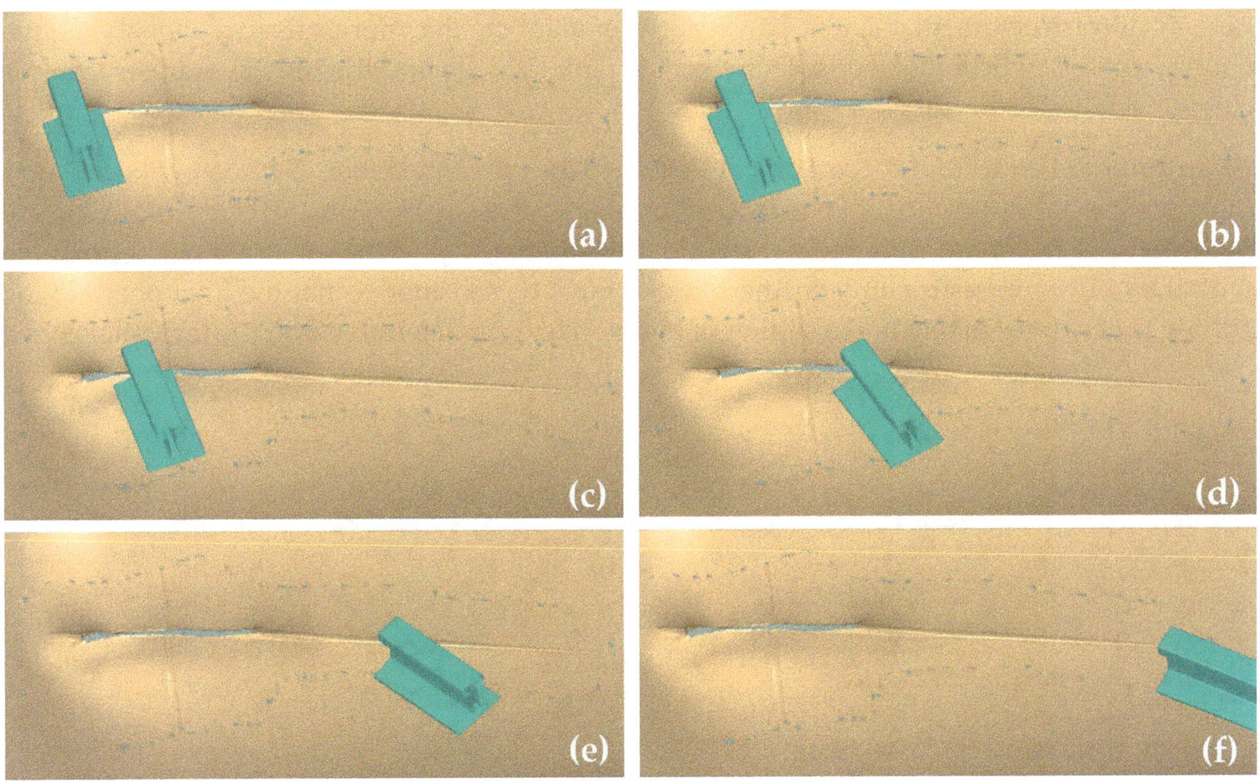

Figure 35. Simulation of the interaction between track reference stake no. 24 and tank shell side of carriage no. 1: (a) initial contact, (b) penetration, (c) start of cutting, (d) end of cutting, (e) start of sliding, and (f) end of sliding.

and track reference stake no. 24 could be geometrically compatible with the shape of the cut on the tank shell.

4.4. Wing rail final exclusion

However, the previous simulation analyses were limited to the verification, from a geometrical point of view, of the local compatibility of the elements possibly responsible for the cut opening with the shape of the cut on the tank shell, without taking into account the positions assumed by the entire tank carriage during the cut generation process.

Therefore, the simulations were newly performed by considering the complete digital models of tank carriage no. 1 (**Figure 37**).

The positions assumed by tank carriage no. 1 during its impact against the elements possibly responsible for the cut were analyzed in the framework of the railway infrastructure with particular reference to the wing rail and track reference stake no. 24 (**Figure 38a** and **b**).

Under the hypothesis of tank shell cutting by the wing rail, **Figure 39** shows diverse visualizations of the positions assumed by the tank carriage no. 1 during the cutting phases corresponding to **Figure 32c** and **33a**.

Figure 40 shows the views of the positions assumed by tank carriage no. 1 under the hypothesis of tank shell cutting by the wing rail corresponding to **Figures 32d** and **33b**.

This simulation evidences a macroscopic interference between tank carriage no. 1 and the railway infrastructure that looks particularly evident in the more advanced phase of the cutting process, where the overlap between the tank carriage and the railway infrastructure is clearly visible (**Figure 40**).

Accordingly, even if from a geometrical point of view, the interaction between wing rail and tank shell is locally compatible with the cut shape, the carriage, in order to allow for the generation of the entire cut opening, should have assumed positions that are not compatible with the railway infrastructure. Furthermore, it must be considered that the final phase of the cutting process, consisting in the sliding of **Figure 32d**, was determined on the tank shell by an

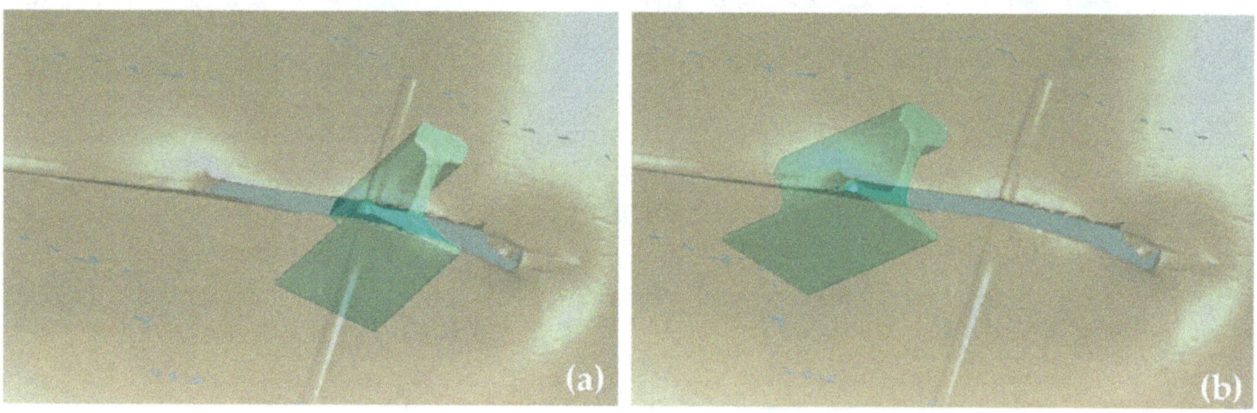

Figure 36. Details of the phases of the cutting action by track reference stake no. 24 corresponding to the position of (a) **Figure 35c** and (b) **Figure 35d**.

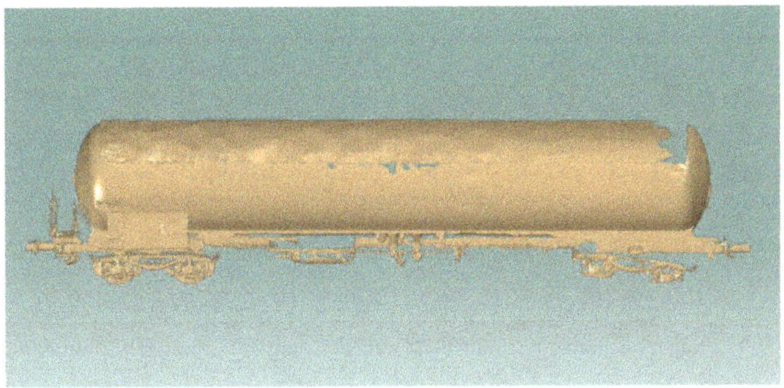

Figure 37. Digital model reconstruction of the entire tank carriage no. 1.

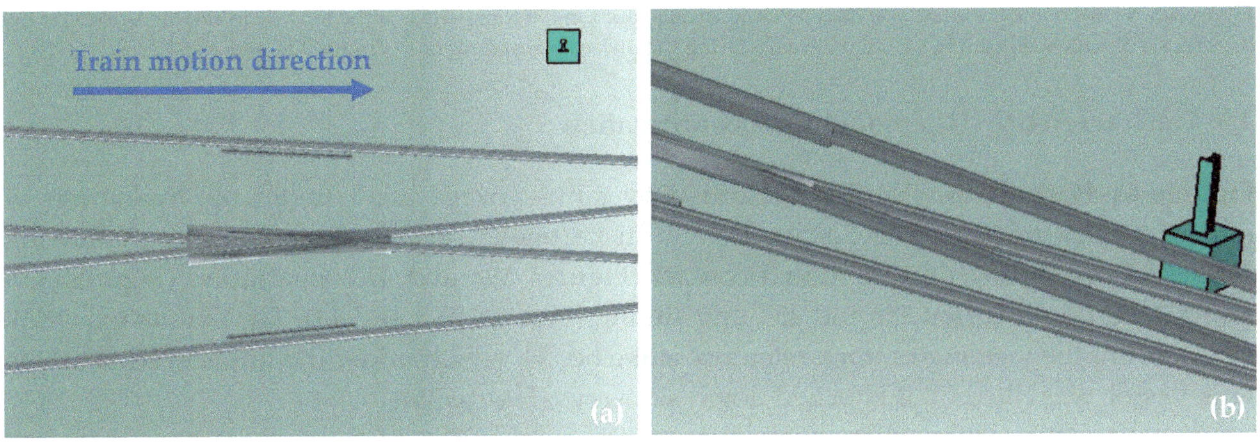

Figure 38. Digital model reconstruction of the railway infrastructure based on the planimetric measurements of the accident scene: (a) position of the wing rail and (b) lateral view of the position of track reference stake no. 24.

object with a sharpened corner shape and not with a blunted corner shape such as the one of the wing rail (**Figure 31**).

It can thus be stated that the wing rail cannot have been the cause of the cut opening on the tank shell side.

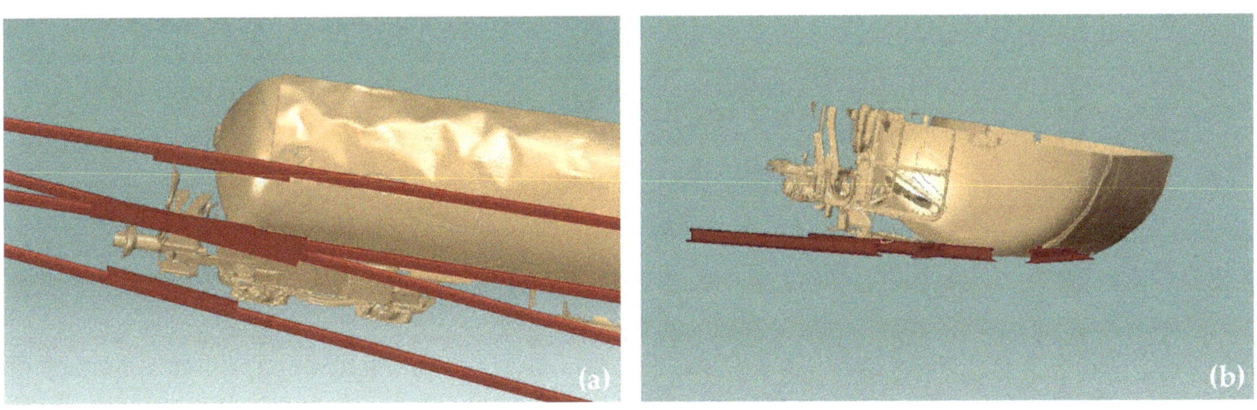

Figure 39. Views of the positions assumed by tank carriage no. 1 in the framework of the railway infrastructure under the hypothesis of tank shell cutting by the wing rail during the phase corresponding to **Figure 32c** and **33a**.

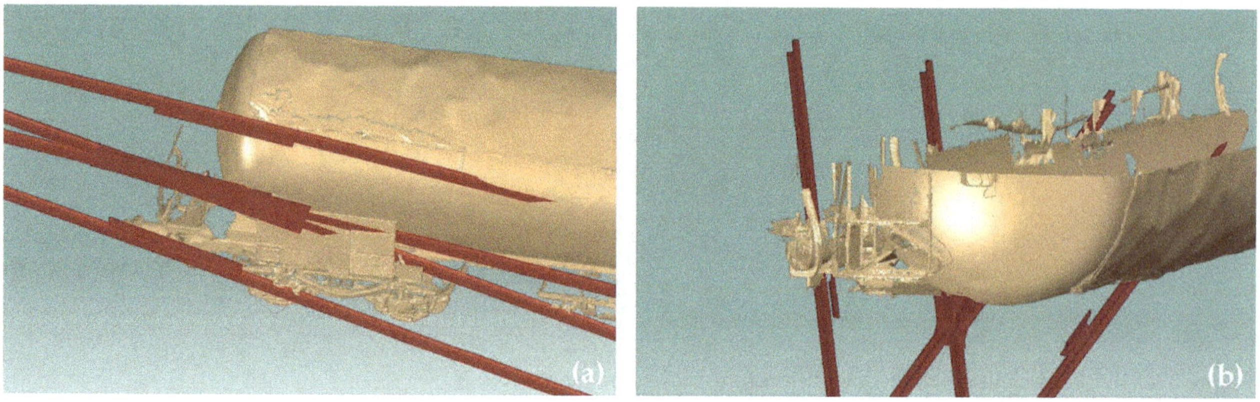

Figure 40. Views of the positions assumed by tank carriage no. 1 in the framework of the railway infrastructure under the hypothesis of tank shell cutting by the wing rail during the phase corresponding to **Figure 32d** and **33b**.

4.5. Track reference stake no. 24 final confirmation

Figures 41-44 show the positions assumed by tank carriage no. 1 under the hypothesis of impact against track reference stake no. 24. In particular, the results refer to the following phases of the cutting process: initial contact (**Figures 35a** and **41**, penetration (**Figures 35b** and **42**), cutting (**Figures 35c** and **43**), and sliding (**Figures 35d** and **44**). The simulations took into account the yielding of track reference stake no. 24, which under the impact action by the tank vessel, was tilted till it knocked down sideways (**Figure 35**).

The simulation results show that, under the hypothesis of tank shell cutting by track reference stake no. 24, no interference is evidenced between tank carriage no. 1 and the railway infrastructure during the tank shell cutting process.

Accordingly, the tank of carriage no. 1, already overturned on its left side, impacted with its front part against track reference stake no. 24, which started the cutting action by penetrating for a certain depth inside the tank shell and continued to cut the sheet metal of the tank shell till it exited from the cut opening after its knocking upon yielding of the ballast where it was grounded.

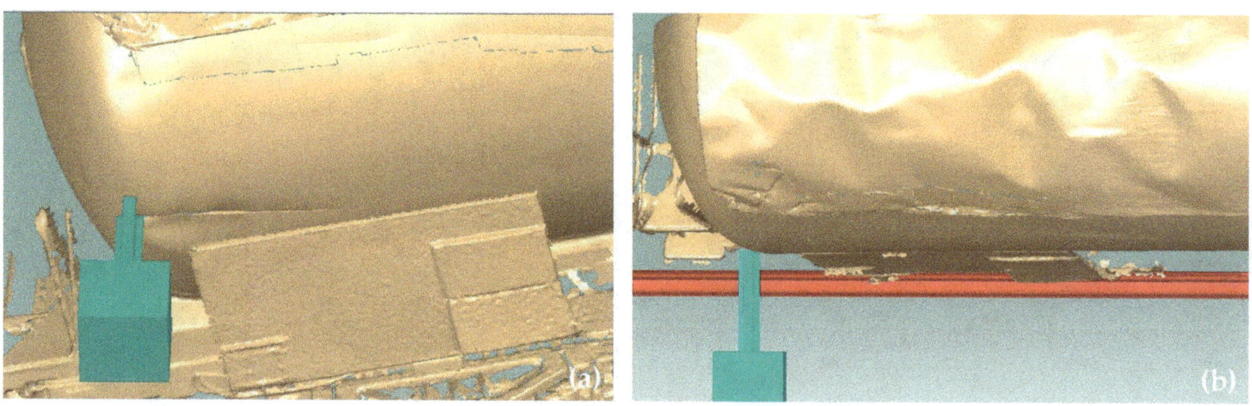

Figure 41. Position assumed by tank carriage no. 1 in the framework of the railway infrastructure under the hypothesis of tank shell cutting by track reference stake no. 24: initial contact phase (**Figure 35a**).

Figure 45a and **b** shows the details of the position of track reference stake no. 24 during the penetration phase into the tank shell.

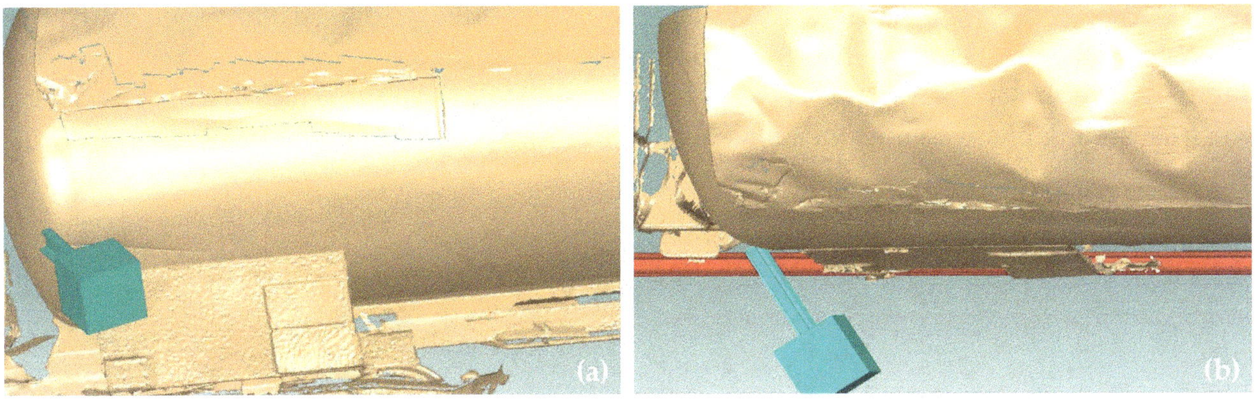

Figure 42. Position assumed by tank carriage no. 1 in the framework of the railway infrastructure under the hypothesis of tank shell cutting by track reference stake no. 24: penetration phase (**Figure 35b**).

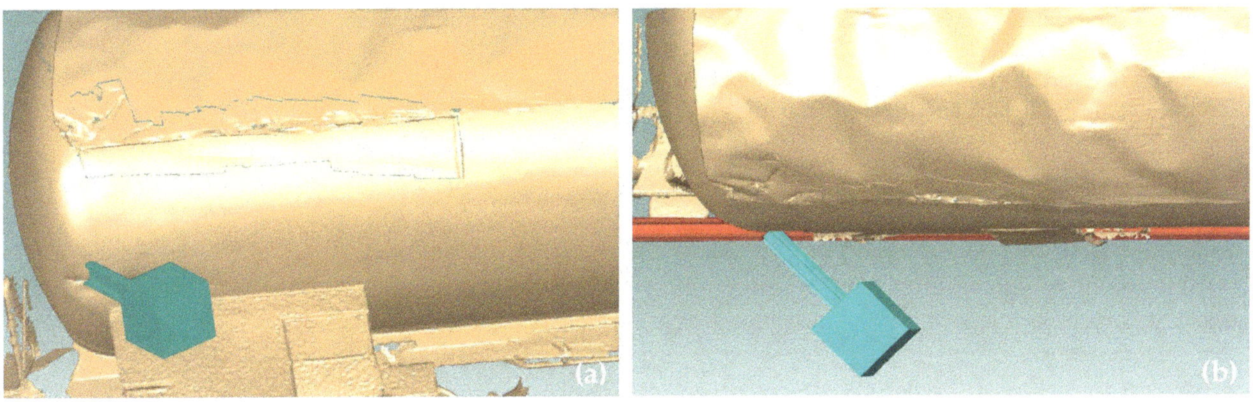

Figure 43. Position assumed by tank carriage no. 1 in the framework of the railway infrastructure under the hypothesis of tank shell cutting by track reference stake no. 24: cutting phase (**Figure 35c**).

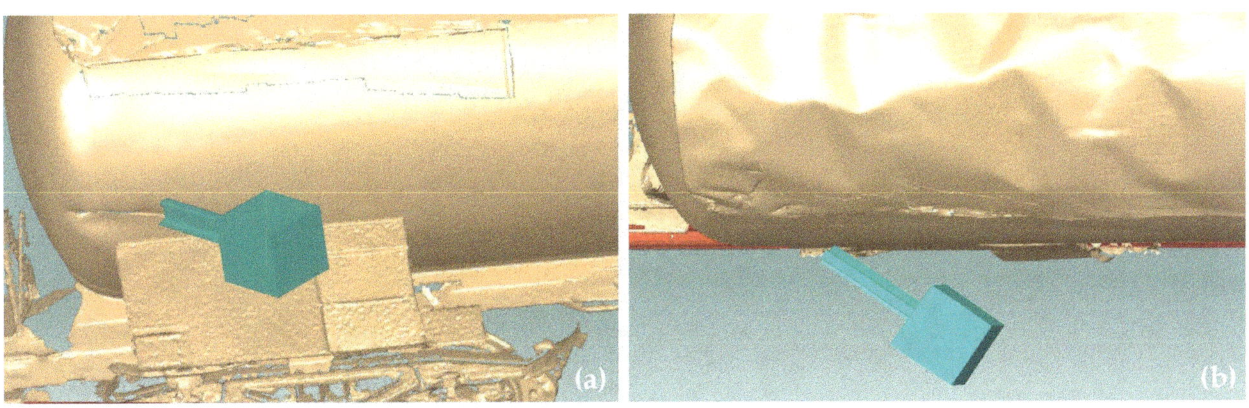

Figure 44. Position assumed by tank carriage no. 1 in the framework of the railway infrastructure under the hypothesis of tank shell cutting by track reference stake no. 24: sliding phase (**Figure 35d**).

During the last phase of interaction, track reference stake no. 24, reversed on the track ballast, could not continue its cutting action on the sheet metal of the tank shell; however, it stuck out of the ground enough to be able to mark the tank shell with a long scratch starting from the end of the cut opening. During this final phase, the sliding between tank shell and track reference stake no. 24 determined the abrasion of the sharp edge of the stake, as evidenced in **Figure 37**. **Figure 46** shows diverse views of the position of track reference stake no. 24 during the final sliding phase against the tank shell.

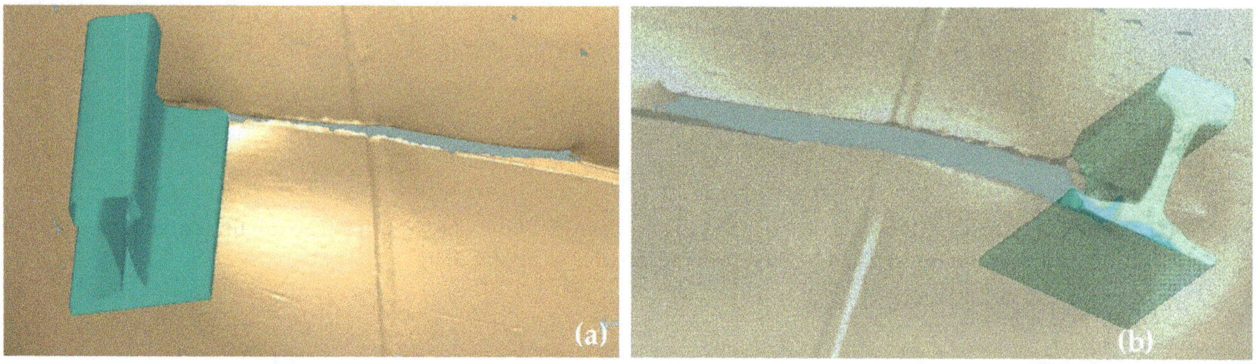

Figure 45. Details of the position of track reference stake no. 24 during the penetration phase into the tank shell: (a) view from the outside of the tank and (b) view from the inside of the tank.

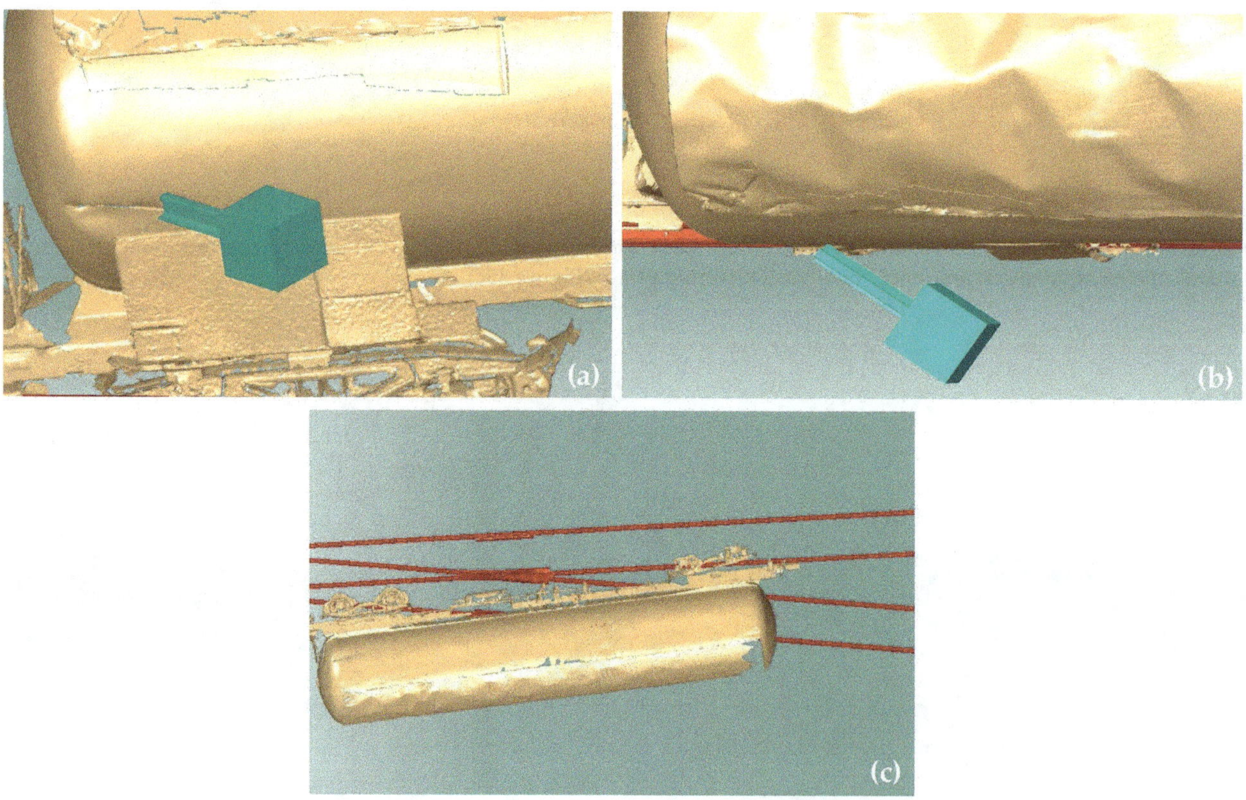

Figure 46. Position of track reference stake no. 24 during the sliding phase against the tank shell.

It can thus be stated that the track reference stake has been the actual cause of the cut opening on the tank shell side.

5. Conclusions

With reference to a real catastrophic train accident where a railway tank carriage, carrying a pressure vessel filled with liquefied petroleum gas, collided with an obstacle that generated a cut in the tank shell, causing a gas leakage with consequent explosion and fire responsible for a high number of human casualties, traditional railway accident reconstruction procedures identified two potential objects accountable for the cutting of the pressure vessel shell—a wing rail and a track reference stake.

In this work, based on digital terrain models and reconstructed models of the railway carriage and the objects possibly responsible for the cut damage on the tank shell, 3D digital simulation scenarios were created, allowing to position and rotate the carriage interactively to detect every possible collision of the pressure vessel with the infrastructure environment. The scope was to investigate whether the form of the cut in the pressure vessel fits to the damage found on the objects potentially accountable for the cutting of the tank shell and whether the interference between object and shell could generate the resulting metal chip, found in the tank, through an interaction similar to a metal cutting operation.

The digital simulation of the developments of the collision between railway tank carriage and infrastructure allowed to observe the potential carriage displacements during the accident from any angular viewpoint. The interference related to collisions between railway tank carriage and the objects in the infrastructure environment potentially responsible for the cut opening was highlighted, allowing to identify the most likely root cause of the catastrophic railway accident.

The analyses of the cut damage process carried out by simulating the relative displacements between the deformed zone of the tank shell and the objects possibly responsible for the cut provided the evidence that two such objects, i.e., a wing rail and a track reference stake, could be geometrically compatible with the shape of the cut.

However, the analyses were limited to the verification, from a geometrical point of view, of the local compatibility between each object and the shape of the cut shape on the tank shell, without taking into account the positions assumed by the entire tank carriage during the cut generation process.

Thus, the simulations were newly performed by considering the complete digital models of the railway tank carriage. The positions assumed by this tank carriage during the simulated impact against the objects possibly responsible for the cut were analyzed with particular reference to the wing rail and the track reference stake in the infrastructure environment.

Under the hypothesis of cut damage determined by the wing rail, the simulation of the positions assumed by the tank carriage during the cutting phases evidences a macroscopic interference between tank carriage and the railway infrastructure that looks particularly evident in the more advanced phase of the cut opening process, where the overlap between the tank carriage and the railway infrastructure is clearly visible.

Accordingly, even if from a geometrical point of view, the interaction between wing rail and tank shell is locally compatible with the cut shape, the carriage, in order to allow for the generation of the entire cut opening, should assume positions that are not compatible with the railway infrastructure. Furthermore, it must be considered that the final phase of the cutting process, consisting in the sliding against the pressure vessel shell, was determined on the tank shell side by an object with a sharpened corner shape and not with a blunted corner shape such as the one of the wing rail. It can thus be stated that the wing rail cannot have been the cause of the cut opening on the tank shell.

Under the hypothesis of cutting caused by the track reference stake, the simulations of the positions assumed by the railway tank carriage during the cutting phases show that no interference is established between tank carriage and the railway infrastructure during the cut opening process of the tank shell.

Accordingly, the tank of the railway carriage, already overturned on its left side, impacted with its front part against the track reference stake. The latter started the cutting action by penetrating for a certain depth inside the tank shell and continued to cut the sheet metal of the tank shell till it exited from the cut opening as it knocked down upon being removed from the ballast where it was grounded.

During the last phase of interaction, the track reference stake, reversed on the track ballast, could not continue its cutting action on the sheet metal of the tank shell; however, it stuck out of the ground enough to be able to mark the tank shell with a long scratch starting from the end of the cut opening. During this final phase, the sliding between tank shell and track reference stake determined the abrasion of the sharp edge of the stake. It can thus be maintained that the track reference stake has been the actual cause of the cut opening on the tank shell side.

As described in Section 3.1, the traditional system for railway track curve geometry periodic checking is based on the installation of reference stakes made of rail chunks grounded in the ballast every 10 m next to the track curve to be checked and protruding vertically out of the ground level. This practice poses a major hazard due to the presence of permanently installed vertical rail chunks characterized by sharp corners that, in case of impact against rail carriages for transportation of dangerous goods or of passengers, can cause considerable damage or even provoke catastrophic events.

The introduction of alternative track positioning and control systems, capable to provide more advanced solutions for safer conditions in the railway infrastructure and operations, is highly desirable to avoid future disastrous accidents of the kind reported in this work. One such alternative system is represented by the railway surveying system already introduced by the national railways in Germany and Norway [12]. Its operation requires the setting, along the track of interest, of topographic reference points, the position of which is previously determined through a precision polygon. Through traditional topographic methods, these points are subsequently referred to by the points materialized through fastening devices, incorporated in the concrete basis of the poles or portals of the electric power line, for the positioning of the measuring instrument. A main advantage of this alternative system over the traditional track positioning and control system is that the track axis is referred to an absolute coordinate system, which allows for the "free" positioning of the measurement station outside the influence of the train movement, without interference with the railway operation. A further major advantage is that there is no more need for track reference stakes: this notably reduces the installation costs, due to the much smaller number of low cost reference points, as well as the complete elimination of sharp corner rail chunks permanently standing in the railway infrastructure, thus providing a vital improvement in the railway operational safety.

Author details

Alessandra Caggiano[1,2]* and Roberto Teti[1,3]

*Address all correspondence to: alessandra.caggiano@unina.it

1 Fraunhofer Joint Laboratory of Excellence on Advanced Production Technology (Fh-J_LEAPT UniNaples), Naples, Italy

2 Department of Industrial Engineering, University of Naples Federico II, Naples, Italy

3 Department of Chemical, Materials and Industrial Production Engineering, University of Naples Federico II, Naples, Italy

References

[1] Ministero delle Infrastrutture e Trasporti. Relazione d'indagine sull'incidente ferroviario di Viareggio. 2 aprile 2014. 84 p

[2] Chandra S. Railway Engineering. India: Oxford University Press, Inc.; 2008

[3] Mundrey JS. Railway Track Engineering. India: Tata McGraw-Hill Education. 2009

[4] Bonnet CF. Practical Railway Engineering. UK: Imperial College Press; 2005

[5] Toni P. Ricostruzione della dinamica dell'incidente ed individuazione delle cause che lo provocarono. Consulenza tecnica di ufficio, Procura della Repubblica presso il Tribunale di Lucca, Italy; 2011

[6] Teti, R. Relazione tecnica di parte: Ricostruzione della dinamica dell'incidente ferroviario di Viareggio del 29.06.2009 e individuazione delle cause del disastro. 2012

[7] Caggiano A, Nele L, Sarno E, Teti R. 3D digital reconfiguration of an automated welding system for a railway manufacturing application. Procedia CIRP. 2014;**25**:39-45

[8] Raja V, Fernandes KJ, editors. Reverse Engineering: An Industrial Perspective. London: Springer-Verlag; 2008. 242 p

[9] Bernard A. Virtual engineering: Methods and tools. Journal of Engineering Manufacture. 2005;**219**(B5):413-422

[10] Barone S, Razionale AV. Relazione tecnica rilievi 3D ed analisi delle condizioni geometriche di danneggiamento del carro cisterna 3380 781 8 210-6 del treno 50325 a seguito dello svio del 29/06/2009. Pisa. 21/10/2010

[11] Segreto T, Caggiano A, D'Addona DM. Assessment of laser-based reverse engineering systems for tangible cultural heritage conservation. International Journal of Computer Integrated Manufacturing. 2013;**26**(9):857-865

[12] Aquilino E, Colonna P, Tragni O. Le innovazioni tecnologiche nei metodi di tracciamento e controllo delle curve ferroviarie. In: Convegno SIIV. (Società italiana di infrastrutture viarie); Cagliari. 1999. pp. 1-18

Permissions

All chapters in this book were first published in MRE, by InTech Open; hereby published with permission under the Creative Commons Attribution License or equivalent. Every chapter published in this book has been scrutinized by our experts. Their significance has been extensively debated. The topics covered herein carry significant findings which will fuel the growth of the discipline. They may even be implemented as practical applications or may be referred to as a beginning point for another development.

The contributors of this book come from diverse backgrounds, making this book a truly international effort. This book will bring forth new frontiers with its revolutionizing research information and detailed analysis of the nascent developments around the world.

We would like to thank all the contributing authors for lending their expertise to make the book truly unique. They have played a crucial role in the development of this book. Without their invaluable contributions this book wouldn't have been possible. They have made vital efforts to compile up to date information on the varied aspects of this subject to make this book a valuable addition to the collection of many professionals and students.

This book was conceptualized with the vision of imparting up-to-date information and advanced data in this field. To ensure the same, a matchless editorial board was set up. Every individual on the board went through rigorous rounds of assessment to prove their worth. After which they invested a large part of their time researching and compiling the most relevant data for our readers.

The editorial board has been involved in producing this book since its inception. They have spent rigorous hours researching and exploring the diverse topics which have resulted in the successful publishing of this book. They have passed on their knowledge of decades through this book. To expedite this challenging task, the publisher supported the team at every step. A small team of assistant editors was also appointed to further simplify the editing procedure and attain best results for the readers.

Apart from the editorial board, the designing team has also invested a significant amount of their time in understanding the subject and creating the most relevant covers. They scrutinized every image to scout for the most suitable representation of the subject and create an appropriate cover for the book.

The publishing team has been an ardent support to the editorial, designing and production team. Their endless efforts to recruit the best for this project, has resulted in the accomplishment of this book. They are a veteran in the field of academics and their pool of knowledge is as vast as their experience in printing. Their expertise and guidance has proved useful at every step. Their uncompromising quality standards have made this book an exceptional effort. Their encouragement from time to time has been an inspiration for everyone.

The publisher and the editorial board hope that this book will prove to be a valuable piece of knowledge for researchers, students, practitioners and scholars across the globe.

List of Contributors

Fouzia Elbahhar and Marc Heddebaut
Univ. Lille Nord de France, Lille, France
IFSTTAR, LEOST, Villeneuve d'Ascq, France

Cesar Briso-Rodríguez and Juan Moreno García-Loygorri
Universidad Politecnica de Madrid, Spain

Lei Zhang
Key Lab of Wireless Sensor Network and Communication, Shanghai Research Center for Wireless Communications, China

Boban Djordjević and Krmac Evelin
Faculty of Maritime Studies and Transport, University of Ljubljana, Slovenia

Mohammed Ali Berawi
Faculty of Engineering, Universitas Indonesia, Indonesia

Davod Poreh, Antonio Iodice, Daniele Riccio and Giuseppe Ruello
Università degli Studi di Napoli "Federico II", Dipartimento di Ingegneria Elettrica e delle Tecnologie dell'Informazione, Napoli, Italy

Ade Ogunsola
Parsons Infrastructure Group, Riyadh, Saudi Arabia

Jon del Olmo, Fernando Garramiola, Javier Poza and Gaizka Almandoz
Electronics and Computing Department, Mondragon Unibertsitatea, Faculty of Engineering, Arrasate-Mondragón, Spain

Muzaffer Metin, Arif Ulu, Mahmut Paksoy and Murat Emre Yücel
Department of Mechanical Engineering, Yildiz Technical University, Turkey

Lin Jing
State Key Laboratory of Traction Power, Southwest Jiaotong University, Chengdu, P.R. China

Alessandra Caggiano
Fraunhofer Joint Laboratory of Excellence on Advanced Production Technology (Fh-J_LEAPT UniNaples), Naples, Italy
Department of Industrial Engineering, University of Naples Federico II, Naples, Italy

Roberto Teti
Fraunhofer Joint Laboratory of Excellence on Advanced Production Technology (Fh-J_LEAPT UniNaples), Naples, Italy
Department of Chemical, Materials and Industrial Production Engineering, University of Naples Federico II, Naples, Italy

Index

T

U

V

www.ingramcontent.com/pod-product-compliance
Lightning Source LLC
LaVergne TN
LVHW060607210926
848485LV00005B/1381